SECOND EDITION

ESTIMATING
FOR HOME BUILDERS

JERRY HOUSEHOLDER and JOHN C. MOUTON

HOME BUILDER PRESS®
National Association of Home Builders
15th and M Streets, NW
Washington, DC 20005

About the Author

Jerry Householder, builder, professor, and author, has many years of experience as a general contractor. He has built over $100 million worth of residential, commercial, and industrial projects throughout the southeastern United States.

The author is Chairman of the Construction Department of Louisiana State University in Baton Rouge. Previous to that appointment, he was the Director of Graduate Studies in Construction Management in the College of Architecture and Urban Studies at Virginia Polytechnic Institute and State University. Dr. Householder teaches courses in planning and scheduling, construction management, and construction law. A professor for over 10 years, the author holds a doctorate in civil engineering from the Georgia Institute of Technology.

Dr. Householder has written *Scheduling for Builders* and coauthored *Basic Construction Management: The Superintendent's Job*, both published by Home Builder Press, National Association of Home Builders. He also has written for the *ASCE Journal of Construction Engineering and Management, Arbitration Journal, Walker's Construction and Estimating*, and other periodicals. Dr. Householder also has presented seminars on building topics at the NAHB Annual Convention.

John C. Mouton is an Associate Professor in the Construction Management Department at California Polytechnic Institute at San Luis Obispo, California. He has a Master's degree in building construction and has taught college-level classes in the building trades for over 15 years.

He has extensive work experience in the building industry. He was a principal and a construction manager for Mouton Construction Corporation of Lafayette, Louisiana, for 5 years. During this time he consulted and provided management services for a variety of residential and commercial projects. Prior to this period, he was the chief estimator for Herlitz Construction Company of Baton Rouge, Louisiana. In this capacity, he prepared competitive estimates, negotiated contracts, prepared change orders, and trained junior estimators.

Prior to his academic career, Mouton was project manager/superintendent of J. B. Mouton and Sons, Inc., of Lafayette, Louisiana.

Acknowledgments

The authors acknowledge Nick Papadopoulos for his assistance in writing Chapter 7, "Computerized Estimating." Papadopoulos is President of Digital Building Resources, a consulting firm in Alexandria, Virginia, that specializes in integrated computer applications. They also appreciate the helpful suggestions provided by (a) Leon Rogers, Chairman, NAHB Business Management Subcommittee on Accounting and Finance, and Assistant Professor of Construction Management, Brigham Young University, Provo, Utah; (b) Steve Watt, Estimating Production Manager, Timberline Software Corporation, Beaverton, Oregon, both of whom reviewed the entire manuscript; and (c) Jean Carmichael, NAHB Software Review Program Coordinator, who reviewed Chapter 7.

This second edition of *Estimating for Home Builders*, was produced under the general direction of Kent Colton, NAHB Executive Vice President, in association with NAHB staff members James E. Johnson, Jr., Staff Vice President, Operations and Information Services; Adrienne Ash, Assistant Staff Vice President, Publishing Services; Rosanne O'Connor, Director of Publications; Doris M. Tennyson, Director, Special Projects/Senior Editor; David Rhodes, Art Director, and Carolyn Poindexter, Editorial Assistant.

Estimating for Home Builders, second edition
ISBN 0-86718-372-1

Printed in the United States of America.

Library of Congress Cataloging-in-Publication Data

Householder, Jerry
Estimating for home builders / Jerry Householder and John C. Mouton.—2nd ed.
p. cm.
Includes bibliographical references.
ISBN 0-86718-372-1

1. Building—Estimates. 2. Building—Cost control. I. Mouton. John C. II. Title.
TH435.H885 1992 91-35895
692'.5—dc20 CIP

For further information, please contact—

Home Builder Press®
National Association of Home Builders
15th and M Streets, NW
Washington, DC 20005

3/92 Scott/McNaughton & Gunn 3.5K

Contents

Figures

Preface

An estimator must understand the estimating process and the relation of the various estimating activities, as well as the practical methods of accomplishing estimating functions. This book describes the process of developing complete construction cost estimates, as well as shortcut methods of estimating, that ensure success in the building business.

Several chapters offer extensive detail about the judgment required for reliable cost estimating, while others specify how to complete individual segments of the residential cost estimate.

Estimating for Builders begins with an overview of the preparation of an estimate, including basic guidelines that aid in completing a rapid, accurate cost estimate. The book details the cost factors that must be included in a complete cost estimate for a residential building project: subcontracts; material; labor; tools, supplies, and equipment; job-site costs; overhead; and markup.

Estimating for Builders discusses the use of certain tools that estimators use. It provides sample checklists and reviews each step of the takeoff process with an actual floor plan to illustrate the process.

A complete cost estimate using a specific example accompanies an explanation of the logic the estimator must use in making the many decisions involved in an estimate. The discussion of the process of checking the accuracy of an estimate includes ways to improve accuracy.

A brief examination of how estimating relates to overall cost control is followed by an investigation of the current status of computer estimating with examples of computer output.

Whether you are a novice estimator or merely want to improve your current estimating procedures, you will find help in the following pages.

Chapter 1

An Overview of the Builder's Estimate

An estimate of a building project is a compilation of the anticipated cost of supplying the materials, labor, subcontracts, and equipment necessary to complete the project. An estimate is valid if it is based on information that has been prepared and completely reviewed according to established procedures. The accuracy of an estimate is directly related to the detail the estimator includes in the estimate.

Before anyone can complete a cost estimate for a house or other structure, he or she must understand construction, takeoff, estimating, and subcontracting terminology. The glossary at the end of this book contributes to such an understanding.

A builder's approach to estimating depends on his or her building operation, market conditions, local practices, and types of clients and contracts. Builders use estimates for scheduling, cash flow, competitive bidding, for purchasing and job cost control, and as a basis for negotiating a project. Even though a builder may get only a portion of the jobs on which he or she bids, estimates prepared for some jobs should be complete. However, a builder who sells homes before constructing them or builds on a speculative basis can also use the estimate for controlling material purchases, subcontracts, and job costs.

A builder who uses the same or similar plans repetitively can reuse the estimated quantities of materials. Comparing these quantities with those on invoices and refining them can make them extremely accurate. However, the builder must estimate the total cost of each house individually to make the necessary adjustments for individual site conditions, foundation requirements, and minor changes. If a builder uses the same subcontractors from job to job, and their prices for the same plans are constant, that builder will save time and effort estimating subcontractors charges. Even though hourly labor costs for the same house will vary from one time to the next, a builder's records will provide a good idea of what the labor cost should be.

If builders choose the plans and determine the specifications, to a large degree they can control the cost of the houses they build. In these cases, they can complete estimates from limited plans and specifications because the details and quality of materials will be whatever the builders choose. If builders standardize their choices, this practice saves estimating time and effort; it also simplifies the subcontract bidding process because subcontractors can use their standard materials and methods.

Preparing an estimate for every job is the profitable way to conduct a building business. If builders want to continue to enjoy seeing projects take shape and grow, they will follow standard business practices, including preparing an estimate for each job.

The Estimator's Role

A successful building business depends on complete, accurate, and consistent cost estimates. The estimator's job is to (a) establish, in advance of construction, an estimated overall cost of a house to be built and (b) develop an accurate listing of the parts and pieces. The estimator must know the building process, all the work items and materials required for each house, and how to apply the cost factors involved in completing each step in the building process.

Even though exercising cost control procedures while a house is being built minimizes waste, loss, and inefficiency, such control, in and of itself, cannot assure a profit on a building project. On an item-by-item basis, the original cost estimate must provide an adequate budget to purchase the materials and work required.

Although some builder-developers make a portion of their annual profit from land development activities, most builders must realize a profit from each house they build to be successful.

Builders basically construct a dwelling under two possible arrangements: they build homes under contract or they build speculative homes. The estimator's role in these two situations can be slightly different. First, a contract job, whether it is a new home or a remodeling job, can be on either a lump-sum basis or a cost-plus basis. Under a lump-sum or fixed-price contract the profit is based directly on the estimator's accuracy. Many contract jobs are built from the owner's plans and specifications, and these plans and specifications often contain materials and procedures with which the builder is unfamiliar.

The estimator must not gloss over the specifications or details in the plans and assume that the materials or the required methods are sufficiently similar to those that are customarily used that they warrant no further investigation.

Example—The problems caused by such assumptions can be seen in a recent case in which the builder contracted to construct more than 20 separate apartment buildings. The plans called for horizontal wood siding. Toward the back of the hundreds of pages of specifications was a single sentence stating that the siding was to be of a particular grade and species. The estimator priced a more commonly used siding. The difference in the cost of the two sidings was enough to pay the estimator's salary for 3 years.

A cost-plus contract with a guaranteed maximum price has essentially the same potential for problems as a lump sum contract. Even if a builder does not guarantee a maximum price, he or she almost invariably provides an estimate of the final cost. Cost overruns on cost-plus jobs rarely do anything to enhance a builder's reputation, and as most builders know, a good reputation is money in the bank.

While estimating a speculative house as accurately as a contract house may not seem necessary, a wise builder will take the time to do so. A speculative house will only sell for what someone is willing to pay for it. Given two sets of plans for houses that have the same market value, one will invariably cost more to build than the other. The smart builder will determine how much a speculative house will cost by estimating it rather than by building it first. He or she will also check what adds value to the house and what does not. Builders sometimes waste money on a speculative house by including features that do not increase its value.

Estimating remodeling work generally takes more skill and know-how than any other kind of estimating. Nothing can really substitute for remodeling experience when an estimator is pricing a remodeling job. Doing remodeling work on a lump-sum basis can be financially rewarding because of the relatively high markup. However, doing remodeling work on a cost-plus basis is the safer way to go. A good remodeling estimator will know how to use allowances and contract contingencies. Properly used, these tools can give a builder the best of both worlds. An allowance is nothing more than a cost-plus item stuck in a lump-sum contract.

The foregoing discussion shows that the role of the estimator can vary according to the type of contact and type of building arrangement. But the main point is that the estimator's overall role in a building business is critical to its success.

Types of Estimates

The amount of information available—such as information about the site and details on the plans and specifications—controls the amount of detail in the estimate. The more detail available, the more accurate the estimate can be. Basically estimators use three types of estimates—complete, component cost estimates, and square-foot estimates. The complete estimate is the most accurate, and the least accurate of the three is the square-foot estimate. An estimate of the cost of a house can combine two or even all three types of estimates, and the accuracy of each depends upon the detail involved. However, all estimates need not be done to the greatest degree of detail possible.

Example—If a potential buyer wants a ballpark figure for purposes of having more detailed plans drawn, usually the builder need not take hours counting every item to be furnished, provided that the builder informs the buyer that the estimate is only a ballpark figure. When a builder anticipates that he or she will not win a contract to build a house, conducting a complete cost estimate may not be justified.

Complete Construction Cost Estimate

The complete estimate (also called quantity takeoff or quantity survey) is based on a total breakdown of the number of work items and materials required with verified prices for the material, labor, equipment, and subcontractors for each and every item. The complete estimate includes the direct cost of material, labor, overhead, and profit. This estimate establishes the budget for building and cost control, the subcontractor information and descriptions, and the materials list for purchase control.

A complete cost estimate requires more time and effort than that required to do a component or a square-foot estimate. However, a complete listing of the quantity and the cost of every item required is the most accurate way to price a job. Chapter 4, "Preparing a Complete Estimate," goes into considerable detail about doing a complete estimate with an in-depth takeoff and pricing example.

Square-Foot Estimate

Square-foot estimating is simply multiplying the square footage of a house by some cost per square foot to arrive at the total cost. This process has been used for years to do everything from giving a ballpark price to determining a firm, hard-dollar bid. Predicting the cost of a house on a cost-per-square-foot basis can range from being a reasonable way to price a house to sheer foolhardiness. If a builder is building every home of virtually the same style and of approximately the same size using the same materials, that builder may be able to track the cost of past houses on a square-foot basis and reasonably predict the cost of future houses. However, to be confident of costs based on square footage a builder must be sure that the conditions in Figure 1-1 are met.

Figure 1-1. Conditions for Using a Square-Foot Estimate

- All site work requirements for the house are known and understood.
- Materials and finishes for the house are the same as those commonly used.
- Cost increases for materials, labor, and subcontractors have been considered.
- Weather conditions will be similar to those upon which the square-foot price was established.
- The cost accounting system on which the estimate is based is accurate and fully operational.

A problem with using square-foot estimates is that many component costs and/or building items are not directly affected by the square footage of a house. One such item is plumbing.

For example, suppose (a) two houses (one 1,300 square feet and one 1,800 square feet) each have two and a half baths and one kitchen and (b) that the principal costs of plumbing, including fixtures and riser piping, are the same for both houses.

Figured on a cost-per-square-foot basis, plumbing component costs are as follows:

$8,600—1,300 sq. ft. = $6.62 per sq. ft.
$8,900—1,800 sq. ft. = $4.94 per sq. ft.

This plumbing component cost estimate for the smaller house is $1.68 higher per square foot than the plumbing component cost estimate for the larger house ($6.62—$4.94 = $1.68).

The use of the square-foot method for estimating the cost of a house also can lead to erroneous results when the builder adds or subtracts floor space.

Example—If the square-foot cost for a house of a particular style and size is $50, a 2,000 square-foot house would cost $100,000. If a potential buyer wants to subtract 100 square feet from the size of the house, at $50 per square foot, supposedly the builder would cut the price $5,000. However, if the bathrooms, the kitchen, the electrical, and the fireplaces all remain the same, would this change actually cut the cost by $5,000? Not likely because the cost of these big-ticket items is spread over the entire area.

Even with substantial historical information, the square-foot or volume method of guesstimating is not recommended for anything other than ballpark pricing, unless the builder continually conducts comprehensive cost evaluations and analyses on many construction projects.

In calculating the cost of a house using the cost-per-square-foot method, the builder must decide what factors to include in the cost per square foot.

Example—Does the price per square foot include the price of the lot, the water and sewage systems, driveways, and walks? The typical way of pricing a house on a square-foot basis is to price the house itself and add the cost of the other items unless they are typically the same for all the houses being built. Some builders like to include their profits in the unit cost while others add their profits separately.

Finally, the builder must determine what compromises the square footage of a house. Certainly a garage costs less per square foot to build than the living area. The covered porches and basements also cost less. Many builders count the living area at 100 percent and add up all the other areas under roof and count them at some reduced percentage, typically 50 percent.

Example—A house with a total area under roof of 3,000 square feet (2,000 square feet of living area and 1,000 square feet of porches and garages) would have an effective area of 2,500 square feet with the house counted at 100 percent and the other areas counted 50 percent:

2,000 sq. ft. @ 100%	=	2,000
1,000 sq. ft. @ 50%	=	500
Total		2,500

Component Cost Estimate

The component cost estimate (also called unit-price estimate) is based on the cost of each part or piece of the entire house. In component cost estimating the estimator divides the project into a limited number of components and estimates the cost of each.

Example—When estimating the cost of a concrete driveway, many estimators know that their cost is "x" dollars and cents per square foot. This cost includes the concrete, the formwork, and finishing. The estimator does not break out these individual items for pricing purposes.

Builders sometimes use short-form estimating procedures when time or information preclude a complete cost estimate. They should proceed with caution when using any short approach to cost estimating, but usually component cost estimating is more accurate than square-foot estimating.

Using an estimate based on component costs instead of square footage gives the builder the ability to adjust the estimate by individual cost and/or by building item as the plans for the house are detailed and completed. Component cost estimating also adapts easily to the use of allowances. A sample estimate for an exterior wall might look like Figure 1-2.

The builder can calculate the cost of the various exterior wall systems normally used in accordance with the varying cost of the materials specified. Likewise, the builder can calculate the cost per lineal foot of the various interior wall arrangements and the cost per square foot of floor area. In addition a square foot of floor area for a slab house would include the cost of a square foot of gravel, vapor barrier, concrete, floor covering, gypsum board, insulation, roof framing system, decking, and roofing. The roofing and decking costs would naturally reflect an increase to account for the degree of slope.

Figure 1-2. Sample Estimate for Exterior Wall

Item	Dollars per Linear Foot
Footings	5.65
Studs and plates	2.21
Gypsum board (painted)	4.30
Base and crown	.60
Insulation	2.40
Sheathing	2.00
Siding	5.00
Frieze	.40
Soffit	.38
Fascia	.50
Roof overhang	1.47
Gutters	1.80
	$26.71

To use the component cost estimating procedures, the estimator simply determines the number of square feet and multiplies it by the cost per square foot of floor, roof, or wall. The number of lineal feet of exterior wall is multiplied by its cost, and the number of feet of interior wall is multiplied by its cost. To the total of these numbers, the estimator adds the cost of the other components such as doors, windows, cabinets, fireplaces, plumbing, electrical, and so on.

sq. ft. of items such as floor and roof x cost per sq. ft.
lin. ft. exterior wall x cost per sq. ft.
lin. ft. interior wall x cost per sq. ft.
+ cost of components (such as windows, doors, and plumbing
total component cost

Figure 1-3 provides a sample component cost estimate for a house with slab-on-grade, hip-roof construction. (A component cost estimate for a house with a basement includes that additional estimating cate-

Figure 1-3. Sample Component Cost Estimate

COMPONENT ESTIMATE
RESIDENTIAL PLAN: SAMPLE HOUSE

COST CATEGORY	AMOUNT
SITEWORK (KNOWN)	$ 2800 -
EXTERIOR WALLS	5700 -
INTERIOR WALLS	3000 -
FLOORS/CLGS./ROOF	14350 -
EXTERIOR DOORS	1300 -
WINDOWS	1900 -
INTERIOR DOORS	1200 -
CABINETS	600 -
FIREPLACE	1500 -
PLUMBING	4200 -
ELECTRICAL	2800 -
FIXTURE ALLOWANCE	NONE -
HEATING/A.C.	2500 -
APPLIANCES	2000 -
JOB OVERHEAD	4900 -
CONTINGENCY/MARKUP	8000 -
TOTAL w/o GARAGE	$56750 -

gory containing the following items: excavation, basement walls, waterproofing, backfill, and floor system.)

The costs included in each category of the estimate are listed in Figure 1-4.

The units used for a component cost estimate are large in scope to simplify the process of determining quantities. The specification and quantity listing identifies both known and assumed items. Estimators would select the component categories that are most consistent with the cost information available.

Example—Some builders use the floor plan's square-footage units for electrical, plumbing, HVAC, and gypsum board. Other estimators forecast by counting the number of electrical fixtures, outlets, switches, and appliances and by measuring the size of air-conditioning and furnace units. Plumbing fixtures, hose bibs, water heater, and sewer and water taps are counted, and the well and the sewage-treatment system, measured.

Figure 1-4. Costs Included in the Estimate

Sitework—Surveying, preliminary layout, clearing and grubbing, rough grading, underground service, clean-up, final grading, landscaping

Exterior Wall—Footings (including excavation), plates and studs, gypsum boards installed and painted, base and crown, insulation, sheathing, exterior finish (such as brick veneer), frieze, soffit (including vents), fascia, roof overhang, gutters and downspouts

Interior Wall—Studs and plates, gypsum board installed and painted (both sides)

Floor, Ceiling, Roof—Soil termite treatment, gravel, vapor barrier and welded wire mesh, finished concrete floor, floor covering, ceiling, roof framing, decking, and roofing.

Exterior and Interior Doors—Count material and hardware.

Windows—Count material.

Cabinets—Count linear foot and note type.

Fireplace—Usually stable lump-sum subcontract price.

Plumbing, Electrical, and Heating, Ventilation, and Air-Conditioning (HVAC)—Vary with equipment and fixture specifications.

Electrical Fixtures and Appliances—Contracted usually on an allowance basis.

Jobsite Overhead—Varies with project.

Contingency—Allowance for the unexpected.

Markup—Reflects competitiveness of the market.

Conceptual Estimating

To review briefly, complete cost estimating is the process of counting all the individual parts of a house and estimating its cost based on the cost of each item. Square-foot estimating involves multiplying the square footage of a house by some cost per square foot. Component cost estimating divides the house into large components, such as exterior walls, interior walls, floors and roof, plumbing, and determining the cost of each component.

All other forms of estimating are simply combinations of these three basic types. Builders commonly combine methods used for the complete estimate and the component cost estimate.

Example—The builder may estimate the framing, drywall, floor covering, and so forth in a detailed manner but price the fireplace, driveway, and other similar components on a component basis.

Conceptual estimating is a process of predicting costs that may use all three of the basic approaches.

Example—A builder may do a detailed estimate of a house, and if the buyer then wants to add or subtract components or area, the builder may determine the changes in the cost by the square foot or component method. If the buyer wants to increase the size of the garage, the builder may know that garages of the type under consideration cost $25 dollars per square foot to build and calculate this increase in cost on a cost-per-square-foot basis. If the buyer wants to know the increases in the cost of adding electrical outlets, a bathroom, or a fireplace, the builder determines the cost by the component method.

Allowances

Contractors are sometimes required to build from incomplete plans and specifications. In these cases the estimator must fill in the blanks in the estimate with allowances. Builders commonly give prices based upon allowances for such items as floor covering, light fixtures, or hardware. For plans and specifications that are even more incomplete, builders may also put an allowance in the estimate for kitchen cabinets, appliances, or some other item that usually is specified. The builder absolutely must qualify such a bid by listing all the allowances so that the client will have a complete understanding of the scope of the proposed work. Because some items can vary greatly in price, the builder should determine the quality of the material that the owner desires for each allowance being given. Knowing the quality desired helps a builder to provide a realistic allowance. An inexperienced builder might be tempted to put in a low number for an allowance in order to entice the buyer to enter into a contract and later try to get the buyer to increase the allowance through a change order. But experienced builders know this practice can be detrimental to a builder's reputation and may be considered to be somewhat unethical.

Overview of a Complete Estimate

The primary purpose of this book is to explain in detail how a complete estimate is done. Square-foot and component cost estimating are explained in general terms, and their application is demonstrated, but the majority of the rest of this book will involve preparing a complete estimate. The steps to be followed are summarized in Figure 1-5.

Familiarization Review

When presented with a new set of plans and specifications, the first thing that the estimator should do is to study the documents thoroughly. The estimator should build the project mentally from start to finish and make notes about those items for which the documents are incomplete.

In this review process, the estimator should look for any aspect of the plans and specifications that might be different from the materials or methods that the builder normally uses, such items as special framing requirements, unusual or unfamiliar materials, as well any special equipment needed.

Even though a topography map may be provided on the plot plan, the estimator should perform a site inspection. While the builder will

Figure 1-5. The Steps to a Complete Estimate

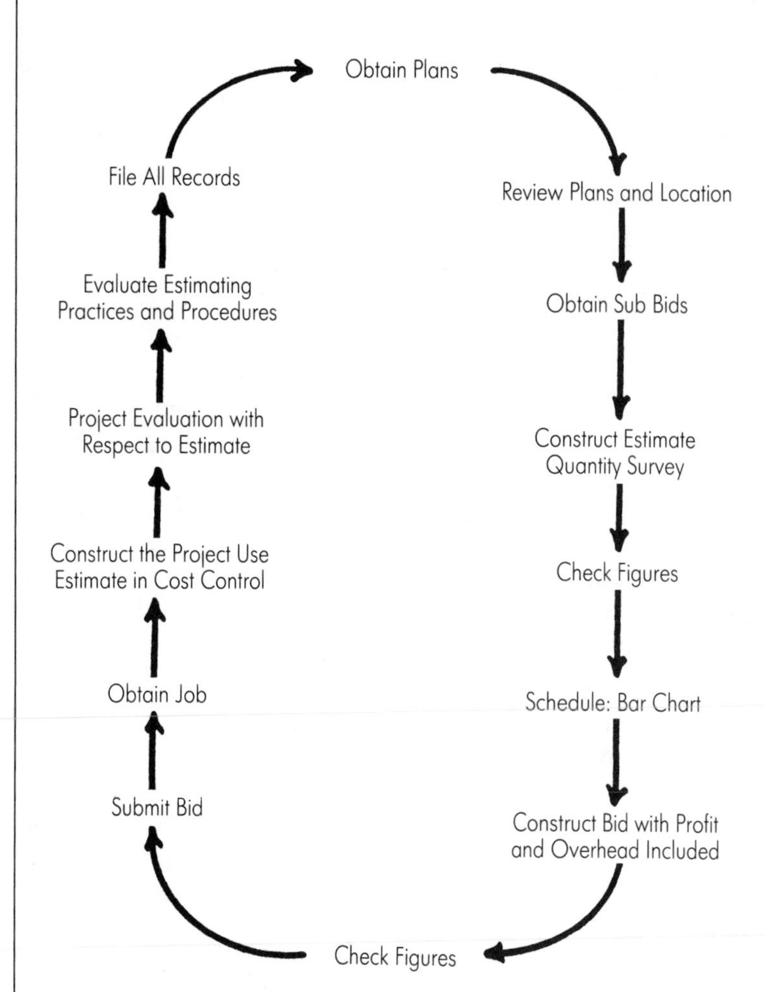

- Review the plans and specifications.
- Inspect the site.
- Request subcontract bids.
- Prepare a quantity takeoff.
- Determine the price of the items to be furnished by the builder.
- Determine job indirect costs.
- Summarize the estimate.
- Determine the profit and general overhead.

Source of Illustration: Leon Rogers, *Residential Construction Estimating*, workbook for a Business Management Certificate Seminar (Washington, D.C.: Home Builders Institute, National Association of Home Builders, 1988), p. 2. Reprinted with permission.

usually not be held responsible for hidden conditions that increase the cost, most courts of law will hold the builder responsible for obvious problems at the site. Some points of special interest to look for at the site are listed in Figure 1-6.

Even though the builder plans to subcontract the rough site grading, he or she should personally inspect the site.

Quantity Takeoff

The quantity takeoff is the listing and quantifying (by category) of all the materials to be furnished by the builder. The best takeoff tool an estimator can use is a good checklist. Each builder should establish his or her own list based upon the customs and practice of the local area and the houses he or she builds. Sample checklists appear in Figures 3-2, 3-3, and 3-15 in Chapter 3, "The Quantity Takeoff Process."

Figure 1-6. Site Inspection Considerations

- Lot corners
- Utility locations including sewer tie-in
- Rock outcroppings
- Unsuitable soil conditions
- Evidence of underground obstructions
- Special access problems
- Unloading and on-site storage areas
- Possible underground water problems
- Environmental impact (especially as it pertains to the everchanging laws on wetlands) and potential hazardous material problems
- Neighbors
- Children

The estimator should be thoroughly familiar with the items on the builder's standard checklist. When reviewing the plans and specifications, the estimator should make a list of items not on the standard list. To prepare the quantity takeoff, the estimator goes through the checklist in an orderly manner and determines the quantity of each item by measuring or counting.

Pricing the Job

The takeoff method demonstrated in this book uses only one set of estimating sheets that combine the quantity takeoff, the pricing, the labor, and the recap. In estimating commercial construction, builders commonly use a more complicated system.

All estimators should ask themselves, "Why guess when you can know? Why guess how much an item will cost when a phone call to the supplier tells the builder for sure? Why guess what a subcontract will be when you can get a firm bid?"

After the estimator does the takeoff of the materials that the builder will need, he or she can list the different items and have local suppliers price them.

The wise builder will always be shopping for good value at the best price. Many suppliers will give special discounts if a builder will buy all the materials at one place. But not all suppliers do. Some builders obtain written quotes from suppliers for the materials commonly used with the agreement that the supplier will not raise the prices without first notifying the builder in writing. Some builders can even tap into their suppliers computer system by modem and obtain up-to-date prices that way.

Figure 4-19 in Chapter 4, "Preparing a Complete Estimate," provides a complete estimate—an actual material quotation for the various items required for a particular house.

In addition to pricing materials, the estimator must also price the labor to be performed by the builder's own labor force and the equipment necessary to do the job. These pricing strategies are discussed in detail in Chapter 3, "The Quantity Takeoff Process," along with subcontractor pricing and overhead costs.

Subcontractor Bids

Most houses can be completely estimated in a relatively short period of time. The process, including the site inspection, usually is a matter of hours. Usually the holdup is subcontractors' bids. Typically the plumbing, electrical, and heating, and air-conditioning take the longest. Getting the plans to the subcontractors early can help to expedite their bids. Many subcontractors such as framers, drywall hangers, and finishers have set prices per square foot or other unit.

Builders can also can save time if they negotiate price agreements with their plumbing, electrical, and heating and air-conditioning subcontractors. Such agreements let builders estimate the subcontractors' work based on some formula such as so much per fixture, outlet, or ton of air-conditioning.

Builders should be sure that all subcontractors bidding on a trade or specialty are estimating the same quality and quantity of work. If the information provided to the subcontractors is incomplete, different bidders may make different assumptions.

Summarizing the Job

After the estimator has priced all of the items and calculated the amount of each category, he or she totals the direct costs. The builder uses the subtotal for each category to create the budget line items for cost control purposes. Sometimes builders base draws or periodic payments on the summarized totals of the various categories as builders in commercial construction usually do.

After figuring the direct cost, the estimator adds the general or home-office overhead and profit to the price to arrive at the bid or sales price. (Home office refers to the builders headquarters office as opposed to a field office not to an office in a home.)

Contract Proposal

The initial purpose of an estimate is to serve as a bid or a proposal to contract to build a house. In most cases the builder will contract to build for one lump-sum amount. Therefore, the builder must be sure (a) to include all costs and (b) that the takeoff and the estimate are accurate because once the bid is accepted, the builder is obligated to contract for that amount. The bid amount fixes the amount the builder will receive to carry out the building contract.

A specific preconstruction sales agreement also fixes the amount to be received. Therefore, the amount of the contract must be based on an accurate forecast of the cost of the required work.

In addition to using estimates to establish the contract prices on lump-sum work, builders also use it to establish budgets for cost-plus work. The accuracy of an estimate for cost-plus work can be just as important as in lump-sum work. A lump-sum or fixed price contract is an arms-length relationship. The builder agrees to furnish the house in accordance with the plans and specifications, and the owner agrees to pay. A cost-plus contract is a fiduciary agreement wherein the builder impliedly if not expressly agrees to use his or her best efforts to keep the owner's best interest at heart. A poorly conceived estimate for a cost-plus job could be viewed as less than the builder's best effort.

Competitive Bid Strategies

All bids are competitive. Builders often bid against other builders. However, in negotiated contracts, builders can be competing with the buyers' budgets. Speculative builders must sell their products in a competitive market, and therefore, they must exercise sound estimating procedures and cost control techniques. However, the most easily understood competitive bid situation is one in which several builders submit bids for a particular house on a certain site, with the contract usually going to the lowest bidder.

The best rule regarding competitive bidding is to bid an amount that includes all direct costs as estimated, a reasonable budget for contingency cost and jobsite overhead, and the minimum markup that satisfies your general overhead and produces a reasonable profit. Therefore, if a builder's bid is successful, he or she can expect a satisfactory margin.

When a builder is bidding competitively on a lump-sum basis against others, he or she should obtain as much information as possible before calculating and submitting a bid. Since material cost estimates and subcontractors' quotes will be similar for most bids submitted, labor, overhead, and markup items will account for the difference in bids.

Likewise, potential cost overruns or underruns arise in those same areas. Contracts are usually awarded to the low bidders (in custom building it is usually to the best reasonable bidder); therefore, to be successful, a builder should not omit or underestimate any cost. On the other hand, including too large a contingency or otherwise overestimating costs may produce a bid that is too high and cause the loss of the contract.

When builders' bids are not accepted, those builders not only have lost the opportunity to profit from building those houses, but they also have incurred the expense of preparing the cost estimates and bids. Because builders cannot expect to win every contract they bid, they should bid only those projects on which they can reasonably compete with other bidders, and thereby, increase their percentage of successful bids.

Succeeding in bidding against builders who underprice their work is difficult. However, these builders usually do not stay in business a long time. Builders should find out who else is bidding before preparing a cost estimate. If builders know their competition, they can more effectively concentrate their estimating and bidding efforts.

In some instances the low bidder may be the builder who has made the greatest error in quantity or figuring costs. Winning because of a mistake in the bid is not winning. Professional builders cannot always compete with bids based on inaccurate cost estimates. A competitive bid is competitive in every cost category: material, subcontracts, labor costs, equipment costs, jobsite overhead, contingency cost, and markup.

A competitive estimate of material costs must include the most competitive price within the quality range required. This price can be determined only after thoroughly collecting material cost information. In addition, a builder should use vendors' discounts for prompt payment in pricing materials if other bidders do. And they should purchase those materials at the available discount.

Builders in the same area usually buy materials from the same or competing vendors. With accurate material quantities and within a reasonable variance, the total material cost estimates for two builders should be the same.

In the case of some materials, such as framing lumber, a material cost estimate might reflect a higher quality (grade of lumber) than is specified if anticipated or estimated labor savings exceeds the extra cost of the higher quality materials. Evaluate this possibility carefully in cost estimating for competitive bids.

Because subcontracts are often a substantial portion of the competitive bid, a builder should submit a bid only when he or she has the most competitive subcontractors' bids. If a builder uses the same subcontractors from job to job, the builder might have to review the subcontractors' estimates with them to ensure that both the builder and the subcontractors will be competitive. In preparing a bid the builder should request bid proposals from as many of the subcontractors submitting bids to other builders as possible. Obtaining low bids from qualified subcontractors who do not submit bids to the competition may gain some edge for that builder.

Strong builder-subcontractor relations result in competitive subcontract bids. Subcontractors who can continually work more efficiently are likely to be motivated to give a bid somewhat lower than those furnished to the competition.

The labor cost estimate is the portion of the bid that is most subject to judgment and interpretation. Wage rates in an area are usually similar, but a builder should expect and receive higher productivity when higher wages are paid.

The labor unit costs used in the bid are based on productivity rates as determined from historical labor cost records. To produce competitive bids, builders should not be overly optimistic in estimating labor productivity and risk the resulting high possibility of labor cost overruns.

Rental rates used in equipment cost estimates must be reasonable if a bid is to be competitive. Some builders lower or eliminate their rental rates to try to gain a competitive edge; although this practice may assist in the competitive aspect of the bid, it can produce a loss of profit when the project is built. Builders should be sure to use realistic rates and usage times when estimating the equipment costs for a bid.

Competitive bidders limit the cost estimate for jobsite overhead to the items for which actual cost must be recovered, they are also careful not to forget any of them.

Historical records of jobsite overhead costs are the best source of estimating information. To complete a construction project within budget, a competitive bid requires an adequate estimate for each item and close control of those costs if the bid is accepted.

Other cost items being equal, a bid is usually awarded to the builder estimating the lowest combined contingency cost and markup. If a builder submits a bid without a contingency cost, any costs that were not accurately estimated will eat into the profit. A bid with a conservative contingency cost is not always successful; a successful bid without contingency cost is not always profitable.

The low bidder is often the bidder with the least markup. A builder should determine the normal markup for his or her bidding market. If this markup exceeds his or her usual or needed markup on projects, then the markup and bid can be competitive. However, if the market's markup is less, that builder either cannot be competitive or will have to accept less net margin on competitively bid projects. This consideration helps to determine whether to bid on a project.

Accurate Bid Proposals

An accurate bid is based on an accurate cost estimate and the other items listed in Figure 1-7. Builders should ensure that their cost estimates are accurate. Otherwise they could place their companies at risk by submitting inexact proposals.

In summary, when estimating for a competitive bid, the estimate should include all the costs that a builder reasonably expects, determined as closely as possible, so that the estimate reflects the best judgement of the final cost. The builder then adds profit accordingly. Any builder who is not the low bidder should remember that to lose a job is better than to waste time and energy building it at a loss.

Figure 1-7. Elements of an Accurate Bid or Proposal

- Accurate material quantities
- Competitive and current material cost quotes and prices
- Complete, competitive subcontractor's bids
- Accurate and complete quantifying of work items
- Realistic estimates of labor productivity and current wage rates
- Adequate allowances
- Satisfactory budgets for equipment, tools, and jobsite overhead
- Contingency amount for potential unknown factors affecting cost
- Reasonable markup for general overhead and profit

Qualified Bids

The most accepted way to qualify bids is to is establish allowances for materials or installations that are not clearly detailed in the plans and specifications. The idea is to establish a total bid amount with a qualification that the bid is based on specific requirements.

Example—A builder might qualify a bid as follows:

> . . . to include carpet cost (material only) not to exceed $14 a square yard; to exclude landscaping, seeding, and other exterior work beyond final grading; all exterior paving to be 4 inches of unreinforced concrete.

The proposal must identify any incomplete cost extensions and all factors assumed in completing the cost estimate. The estimator would determine these factors in a manner that is consistent with the portions of plans and specifications that are known. Qualification of the resulting bid helps to prevent disputes during the building of the house.

Many builders find using incomplete plans and qualifying bids to their advantage. By filling out their own "specification of work" and using cost allowances to qualify bids, builders can select and suggest the materials and details that they want. To qualify an estimate prepared from incomplete plans an estimator should attach a list of materials or outline of specifications to the bid. The list also should include a statement of material specifications, required descriptions and dimensions, such as roof size, shape, and slope, and specific details on siding, soffits, and roof overhang. Comprehensive statements of material specifications assure that a bid cannot be misinterpreted, even if the estimate is derived from only a floor plan.

The Estimate for Cost Control

The cost estimate is the budget for building a house. For a builder to achieve the expected profit, the actual construction cost must be less than or equal to the estimated cost. Because a builder's cost control system produces historical cost records, those records become an important source of cost estimating information. Because the budget is derived from the estimate, the level of detail in the cost estimate affects the level of detail the cost control system can provide. The historical data produced by the cost control procedure must be updated to account for inflation and other factors that affect the costs.

To prepare valid cost estimates, an estimator must know and understand the cost control process and system, which is described in detail in Chapter 6, "The Cost Control System."

Estimating and the Construction Schedule

The building process proves that time is money. The estimate of costs and the control of costs include scheduling considerations. The cost categories that are most affected by construction time are labor, equipment, and jobsite overhead costs. Labor unit costs used in the estimate are actually an allocation of time to complete the work with a certain size crew and cost per hour.

Estimates usually list the equipment and the quantities of many jobsite cost items in units of time (hours, days, weeks, months). Any variance in the time actually required from the time budgeted in the estimate produces a cost overrun or underrun.

To accurately estimate the cost incurred in building requires a schedule for the work being done. For the most part, historical records can provide an estimate of the time required to complete individual segments of work or to build the complete house. However, each estimated cost must correspond to a reasonable schedule.

Example—A builder knows from historical records that a certain size house can be framed by a particular size crew at $400 per day in 3 weeks (15 days). The total labor estimate for this segment of work must be at least, and not substantially more than, $6,000 (15 × $400), if the cost estimate is to match the schedule.

Because scheduling work to be done on overtime affects actual cost, overtime should be included in the cost estimate. If the cost estimate for jobsite overhead is based on a duration of 5 months, a cost overrun occurs if building time exceeds this figure.

Scheduling and time management are addressed in the publication *Scheduling for Builders*, published by the Home Builder Press of the National Association of Home Builders.[1] Although building projects are usually scheduled in detail after the cost estimate is complete, the builder may prepare a preliminary schedule during the estimating process. For cost control and estimate control information to be consistent, the preliminary schedule must become the basis for the actual time control schedule.

Aside from materials and subcontracted items, a builder's cost risks are related to time required to complete the work. Therefore, to accurately estimate cost requires scheduling considerations.

Chapter 2

The Complete Estimate

Most builders purchase the various items for the houses they build in a fairly consistent manner, and their takeoffs and estimates should be as consistent as their building processes. For each item listed on their takeoffs, builders incur one or more costs in the construction of a house. The components of direct construction cost include subcontracts, materials, labor, and equipment. This chapter discusses each of these at length.

The cost estimate forms the basis for the budget for buying the items required to construct a house. Therefore, more detail in the cost estimate allows more cost control.

Many builders subcontract for much of the work required for the construction of the houses they build on a unit-cost basis, (at so much a square foot, so much a brick, and so on). In such cases the builder must know that the estimate of the quantity of work is accurate to be confident of the resulting total labor cost estimate.

Some builders buy most of the materials for the houses they construct and employ their own workers to install them. However, the majority of builders buy only certain materials because they subcontract most or all of the labor required to build a house, and the subcontractors buy the materials they use. These builders hire only a few workers on an hourly basis to do pick up work or none at all. Although many builders prefer to control the flow of material to the jobsite and the payments for the materials, who buys what materials usually follows local standard practice.

To prepare a complete material cost estimate, an estimator will need an accurate quantity takeoff and up-to-date quotes from reputable material suppliers for all materials that the builder will furnish.

Although builders should organize their estimating procedures to suit their individual work habits and experiences, the first step in preparing a complete and accurate cost estimate requires that they do the following:

- Identify all the material that goes into the house.
- Determine a method for buying each piece or part.
- Make a complete list and summary of the cost of all parts and pieces (material items).

Many ways exist to purchase the materials for a house, but purchasing and estimating formats must be consistent. Most estimators prefer to arrange their quantity take-offs and cost estimates according to the sequence of construction, and the authors recommend this method. The benefits of this arrangement are as follows:

- "Building" the house on paper helps you to understand all of the costs involved.
- It provides for direct and simple combination with the project schedule for cost control and cash flow.
- The materials list follows the sequence in which materials are needed for construction. Items with long lead times for delivery can be noted and ordered early to meet the schedule for completing the house.
- It helps to ensure that the you do not leave anything out of the estimate.

You should understand all of the cost factors involved in the complete job before attempting to prepare the quantity takeoffs and to estimate the costs of individual work items. Factors you must understand include lump-sum and unit-cost subcontract bids, labor-only subcontracts, writing subcontracts, and soliciting, evaluating, and selecting bids.

Subcontract Bidding

Builders depend upon subcontractors to perform larger and larger portions of the work. In the past, most builders did a majority of the work with their own crews. In those days builders' reputations depended largely upon the skills of their craftspeople.

Several reasons have prompted the increased use of subcontractors. This method of operation gives builders more flexibility. The building industry has its ups and downs. Homebuilding is especially sensitive to swings in the economy. Slight advances in the mortgage interest rate drive many potential buyers from the marketplace. Homebuilding is so sensitive to economic trends that economists look to housing starts as an indicator of how the economy is doing.

If builders are doing all the labor with their own crews, they each have a set output. Because hiring and firing people is a complicated, time-consuming process, this arrangement provides little or no opportunity to quickly expand or contract the business in response to market trends. Subcontracting affords builders that opportunity.

To the extent possible, builders should standardize their approach to using subcontractors and bidding the subcontracts.

A sample subcontract bid summary appears in Figure 2-1. (Other forms to ease the subcontract bidding process appear in Chapter 4, "Preparing a Complete Estimate.")

As trades and crafts become more specialized, providing full-time, year-round employment for specialty craftspeople becomes more difficult for a builder. Consequently, better-skilled craftspeople often work for specialty subcontractors. In many cases the best workmanship is available through subcontractors. However, some areas lack qualified subcontractors for particular trades. This situation requires

Figure 2-1. Sample Subcontract Bid Summary

SUBCONTRACT BID SUMMARY

	CATEGORY CONTRACTOR	1	2	3	4 LOW BID	5	6
1							
2	GRADING AND EXCAVATION						
3							
4							
5	CONCRETE PLACEMENT/FINISHING						
6							
7							
8	FRAMING - LABOR ONLY						
9							
10							
11	MASONRY - LABOR ONLY						
12							
13							
14	PLUMBING						
15							
16							
17	ROOFING - LABOR ONLY						
18							
19							
20	ELECTRICAL - EXCLUDES FIXTURES						
21							
22							
23	HVAC						
24							
25							
26	INSULATION						
27							
28							
29							
30	DRYWALL - INCLUDES TAPE/FLOAT						
31	EXCLUDES GYP BD						
32							
33							
34	TRIM CARPENTRY - LABOR ONLY						
35							
36							
37	PAINTING - INCLUDES PAINT						
38	EXCLUDES WALL PAPER						
39							
40	LANDSCAPE						

builders to include the cost of the work to be performed by their crews or the cost to import highly specialized craftspeople if the builders' own crew is not trained in these tasks (such as creating custom staircases).

For these and other reasons most builders now subcontract a substantial portion of the construction to specialty trade contractors. Some builders subcontract all aspects of the work other than supervision. Usually, a builder chooses whether or not to subcontract a certain portion of the house based on the quality of workmanship desired and the prices quoted by the subcontractors. (For details concerning the decision to subcontract, refer to *How to Hire and Supervise Subcontractors*, published by the Home Builder Press of the National Association of Home Builders.)[1]

Specialty trade contractors are in business to make a profit. The cost of subcontracted work can be somewhat more expensive than similar work performed by a builder's own crew, but not always. A builder's crew usually consists primarily of framing and trim carpenters, and they may have difficulty competing with drywall hangers or roofers in their specialties.

In most instances electrical, plumbing, and heating, ventilation, and air-conditioning (HVAC) work are subcontracted. Most often, all of the HVAC work is subcontracted to one mechanical contractor. Such subcontractors may do all of the work with their own forces, or they may parcel portions of the work, such as ductwork, to other, more specialized contractors.

Some subcontractors provide both plumbing and HVAC trades and will bid for and provide both. A builder also may contract separately for well drilling and equipment, sanitary treatment systems, sewer/water taps, and related exterior plumbing work.

Although electrical work is also usually subcontracted, builders often furnish and, therefore, estimate the lighting fixtures on an allowance basis. Subcontractors' bids and contracts may include the cost of labor for installation of the fixtures, but usually the builders buy and furnish the electrical fixtures based on either allowances or specified fixtures.

Some builders employ carpenters; however, various portions of the carpentry or other work, including formwork, roofing, and sheetrock work may be subcontracted on some or all houses.

Lump-Sum Subcontract Bids

Subcontractors have three basic ways to determine how much they are to be paid for their work—lump sum, unit price, and hourly. A lump-sum bid by a subcontractor is an offer to do all of the work required under the subcontract for one total. Obtaining a lump-sum bid from a subcontractor requires a complete set of plans and specifications for the work to be subcontracted. If only a portion of that information is available, the builder and subcontractor must agree on the scope of the work assumed by the subcontractor when preparing the bid. The subcontractor's bid is qualified by those assumptions.

When subcontracting a portion of the construction on a lump-sum basis, the cost is based on what the subcontractor thinks will be required to complete that portion of the work. A builder must be certain he or she and the subcontractor agree on this part of the project before the bid is accepted and work begins. (See "Writing Subcontracts" in this chapter.)

Unit-Cost Bids

Many builders subcontract parts of the work on a unit-cost basis, especially for labor-only subcontracts. On all unit-cost bids from subcontractors, builders must prepare an accurate takeoff of the number of units to complete an accurate cost estimate. In some instances the measurement of units for a subcontract is direct.

Example—Bricklayers often quote their labor at some number such as $375 per 1,000 bricks or concrete blocks as $1 for each 8-inch standard concrete block. In this case the builder must have an accurate count of the masonry units (blocks or bricks) to determine the cost of the work the subcontract covers.

The total square footage of the house is another indirect cost method often used in subcontracts for framing. Because the subcontractor could use different area computations for the same house, the builder and the subcontractor must agree on the method of area measurement and computation before completing a cost estimate.

Builders often contract for drywall on a square-foot basis, but the standard for payment is usually the total number of square feet in the boards cut up for use, a figure greater than the square footage that the subcontractor hangs.

Because the payment of a unit-cost subcontract is based on counting or measuring (the methods for which vary considerably), the builder and the subcontractor must agree on a standard method of measurement for each of the various work items. This agreement ensures that the builder's estimated quantity equals the subcontractor's quantity.

You must be familiar with the units used in unit-cost subcontract bidding for each subcontracted item. In doing quantity takeoffs you must list each of the work items to be contracted on a unit-cost basis in the appropriate units for figuring the cost of those items.

Hourly Basis Bids

Some subcontractors price their work in units that are not related to the specific plans or specifications. Other subcontractors in the same trade may use different units for costs.

Example—When ABC Excavation Company charges $45 an hour for a backhoe with an operator, the builder must make an accurate estimate of the time required for the work to accurately estimate its cost. If XYZ Excavation Company charges $1 a linear foot for footing excavation and the residence to be constructed has 420 linear feet of footing, the excavation should take 2 workdays. The estimate of the cost of each subcontractor's work is as follows:

$$\text{ABC} = 2 \text{ days} \times 8 \text{ hrs. a day} \times \$25 = \$400$$
$$\text{XYZ} = 420' \times \$1 = \$420$$

Therefore, if the estimate of 2 days is correct, ABC's bid is less. If an extra day is required, the total excavation cost using ABC will be $600. Although XYZ's price is slightly higher, it is fixed by the amount of work required and may be more economical if the time for completion is in doubt.

Other subcontractors may also want to work on an hourly basis, especially for remodeling work or other work for which the time to do the job is uncertain.

When builders pay a subcontractor by the hour, they must make sure that the subcontractor and subcontractor's employees are not classified as the builder's employees because of the legal liability and taxes involved.

An employer is responsible for the acts of an employee insofar as those acts are related to the employee's job. If a builder's employee is going for supplies in his or her own vehicle and injures a pedestrian, the builder will in all likelihood be liable for the damages too. However, if an independent subcontractor injures a person under the same circumstances, the builder probably will not share the liability.

A builder can also incur additional problems with governmental agencies if a subcontractor is reclassified as an employee. If this change occurs, the builder will be responsible for paying the worker's Unemployment Compensation tax and the employer's share of the Social Security tax. If the subcontractor has been paid in full and is no longer working for the builder, the possibility of recovering this expense is remote.

How, then, can a builder make sure that an hourly subcontractor is truly an independent contractor and not an employee? Builders have no absolutely sure way to determine this status. The test that the courts have used to determine a worker's status is whether or not the builder has control over the method of the worker's performance beyond an agreed upon result to be accomplished. If the builder exercises direct control over the worker's actions, the worker is an employee. If the builder asks for a result, does not exercise control, and the worker has latitude about how to achieve the result, that worker is likely to be classified as an independent contractor.

The nature of the relationship between the builder and the worker also determines whether or not a worker is an independent contractor. If the worker works only for the builder and on a continuing basis, day in and day out, the courts and the tax officials will probably classify an hourly paid worker as an employee. However, they would more easily classify another worker doing a one-time project under the same amount of control as an independent contractor.

Labor-Only Subcontracts

In addition to the type of contract (i.e. lump-sum, cost plus, or hourly), the estimator should be knowledgeable about other aspects of the builder-subcontractor relationship. Many trade subcontractors lack sufficient finances to carry the cost of materials in their subcontracts. If builders are concerned about this situation, they can protect their construction projects from material liens by buying the materials themselves. Often, they can buy on better terms and conditions than can the installing subcontractors. This arrangement affords the builders control of material delivery and handling at the job site. However, for some trades such as electrical and plumbing, this arrangement is usually impractical.

A labor-only subcontract is more likely to be subject to problems of waste or loss of materials. A labor-only subcontractor customarily pays craftspeople on an hourly basis, and the builder usually pays the subcontractor on a lump-sum or unit-cost basis. Assuming payment of similar wage rates, the low-bidding subcontractor must get maximum productivity from his or her workers to be profitable. Therefore, speed is critical to profitability. Often, increased speed results in increased waste because of inefficient use of material.

If prices from the subcontractors of a particular trade seem high, builders might want to compare the cost of using their own labor to do the work with the cost listed in the subcontractor's labor bid. (See the section on the "Labor Cost Estimate" in this chapter for information on comparing labor estimates.)

Labor and Material Subcontracts

Subcontracts for major work items (including electrical, plumbing, HVAC, and floor covering) are most often based on labor and material, and they include all tools, equipment, and supplies required to complete the work. To ensure that a subcontractor's bid accurately reflects the actual work required, the builder must provide adequate plans and specifications to the subcontractor. Obviously, specialty trade contractors can prepare more accurate cost estimates when they are bidding on the specific terms and conditions, and accurate subcontract bids are always to the builder's benefit.

For example, the plumbing specification for a house with 2½ baths, a kitchen, and a washer connection can vary greatly because of the following:

- Quality of the fixtures
- Location of fixtures and rooms
- Location of water-supply tap and sewer tap
- Specific local codes

Without complete information, a builder could get substantially different bids from different subcontractors who make different assumptions: the low bidder might be the one who has assumed the lowest quality or least amount of work. But the typical reaction to incomplete information is for the subcontractor to bid the job on the basis of the worst-case scenario.

Bid Evaluation and Selection

One of builder's worst problems occurs when a builder contracts for a portion of the work with a specialty subcontractor at a price that is too low for the subcontractor to adequately complete the work. The subcontractor may cut corners, ask for unjustified extras, and have difficulty completing the contract or paying suppliers and craftspeople. The builder usually ends up paying increased costs as well as expending extra effort to resolve the resulting problems. Such situations confirm the old saying in the construction business, "The lowest price isn't always the lowest price."

However, accepting subcontractors' bids that are too high creates difficulties in staying competitive while making a reasonable profit.

Keeping abreast of the value of the individual items of work to be subcontracted helps builders to evaluate subcontractors' bids. If they are too expensive, builders might elect to reestimate the work and have their own crews do it. But builders must make such decisions carefully because some crafts require such a high degree of specialized skill. The depth and breadth of the skills of a builder's staff would strongly govern such decisions.

To be confident of an estimate, the builder must evaluate the individual bids for each subcontract item before accepting it for use in the cost estimate. Unless the builder and the subcontractor have worked together many times, the builder should get at least three bids for each proposed subcontract. This practice helps identify and eliminate bids that are either too high or too low. Two bids, when close, can provide the information needed for subcontractor selection. A wide cost difference between two bids can make selection difficult.

Example—bids from two companies for the labor to install a slate roof, a nontypical item, are as follows:

Quality Roofing	$6,400
American Roofers	$11,500

Obviously, both bids cannot be accurate unless American Roofers works at an extremely high profit margin. Possible solutions to this dilemma are listed below:

- Ask both bidders (or the lower one) to check their bids for a possible mistake.
- Check a national cost reference guide for the average value of such work.
- Get another bid from a third roofer

Only in the most competitive circumstances should the selection of a subcontractor be made only on the basis of lowest cost alone. Being responsible for the quality of all the work and for completing the house on time requires that a builder evaluate each bid in light of the previous performance of the subcontractor along with other criteria, such as the size and complexity of the work required, the current workload, and the bidder's financial and manpower capacity.

Writing Subcontracts

The description of the work covered by the subcontract should be written before bids are solicited. If the description of work in the subcontract differs in any way from the bid information, the low-bidding subcontractor may want to increase or withdraw his or her bid.

Upon receiving a written quotation from a subcontractor, a builder should examine the terms and conditions of the offer, including any qualifications. Subcontractors often submit bids on their own forms, but regardless of whose form they use, the builder must carefully compare each quote to the listed description of work.

Because most builders subcontract the same parts or pieces of each house that they build, they can write general descriptions of work for the various subcontracts and then specify the particular items that are to be included in, or excluded from, each subcontractor's bid and the subcontract.

The description of work must state that all the labor required for the work items is to be provided by the subcontractor, and it should state which materials, supplies, and equipment for the work are also to be furnished by the subcontractor. If the subcontractor is to provide only some of the materials, supplies, or equipment, the description of work should list the items to be provided.

The subcontract often refers to the drawings and specifications that pertain to the subcontracted work. A sample description of work might read as follows:

> All labor, materials, supplies, and equipment, including piping, fittings, fixtures, and water/sewer taps as shown on the plumbing plan, site plan, and floor plan and described in the specifications. Tap fees are to be paid by the builder. Compliance with local building codes is required.

The subcontract should specify who is responsible for the cost of clean-up and removal of the subcontractor's trash and include a provision for the subcontractor's use of the builder's equipment. Some builders prefer for their own crews to do all miscellaneous work and to be responsible for material handling and equipment because that practice is less expensive than subcontracting such work. Builders who choose this method should be certain they are credited on bids submitted.

The contract should state that the subcontractor is an independent contractor and not an employee. It also should clearly indicate that the subcontractor will be responsible for his or her own federal and state taxes, including Social Security, and Unemployment and Workers' Compensation insurance.

When the builder has a contract with an owner, the subcontracts should have a flow-down clause. Such a clause simply says that the subcontractor is bound to the builder in the same way that the builder is bound to the owner in the builder-owner contract. Without a flow-down clause, the owner can sometimes require the builder to do certain things but the builder cannot require the same performance of the subcontractor. A flow-down clause is not a one-way street. The subcontractor has the same rights against the builder with regard to making claims and protests that the owner-builder agreement gives the builder. This arrangement is viewed by most interested parties as a fair trade-off.

The subcontract should specify when the builder will pay for the work covered by the subcontract. Different options include the day the work is finished, the end of the week after the work is finished, the end of the month, or when the builder gets paid. If the subcontractor is to be paid when the owner pays the builder, what happens if the owner files for bankruptcy and never pays the builder? The answer is that the builder owes the subcontractor unless the subcontract specifically says that if the owner becomes insolvent the builder does not owe the subcontractor. The courts generally hold that the builder is in a better position to judge the solvency of the owner than a subcontractor.

The builder must have the right to terminate a subcontractor under certain circumstances. Common contract language gives the builder the right to terminate the contract 7 days after written notice if the subcontractor repeatedly fails to carry out the work in accordance with the agreement and the plans and specifications.

A builder's contract with the owner does not give the owner an implied right to make changes in the plans and specifications. Therefore, most construction contracts give the owner that right through a change order clause. Even if a contract is silent with regard to changes, a prudent builder will weigh the damage to his or her reputation before refusing to change the work. The agreement between the builder and the subcontractor also does not automatically give the builder the right to make changes in the absence of a written change clause. A change clause should give the builder the right at any time, without invalidating the subcontract, to make changes in the work. The clause should state that the subcontractor has a duty to perform the change as directed and that, if the builder and the subcontractor cannot agree upon a price for the change, the change will be done on a cost-plus basis with the subcontractor responsible for documenting time and materials.

The preceding paragraphs discuss only some of the items to be included in a contract between a builder and a subcontractor. Builders should seek the advice of their attorneys to draft contracts that will best protect them in their particular jurisdictions. (See *Builder's Guide to Contracts and Liability*, published by Home Builder Press of the National Association of Home Builders.)[2]

Soliciting Bids

In preparation for soliciting bids from subcontractors, builders would be wise to follow the steps in Figure 2-2 before contacting specialty trade contractors and requesting bids:

For a builder who works in a limited geographical area, the list of potential subcontractors for each item will be consistent. He or she will be contacting essentially the same bidders for each house. Stipulating a specific bid due date (and time), as well as the format (phone or written) for bids, allows the builder the time and opportunity to evaluate and select the subcontractors' bids before the builder's bid is due for delivery to the customer. Figure 2-3 is a sample invitation to bid. Telephone calls in response to the invitation should be directed to the person responsible for bidding the work.

Figure 2-4 is a sample postcard form for soliciting subcontractors' bid information for those who prefer this method.

Figure 2-2. Preparation for Soliciting Subcontractors' Bids

- List the work items to be subcontracted.
- Write a description of the work for each item.
- List (three or more) subcontractors for each item.
- Establish a bidding format (phone and/or letter by fax or mail).
- Set a deadline (time and date) for subcontract bidding.

Figure 2-3. Sample Invitation to Bid

Project______________________________ Date__________

Subcontractor__________________________ Time__________

Contact____________________ Phone (____)__________

Your firm is invited to bid on

(description of work)

(description of project)

If you are interested, submit your bid by ______(time and date)______.

We will meet with the owner/designer on ______(time and date)______ and will get answers to any questions submitted prior to this date.

Option 1: When do you want to come to our office to use the plans and specifications? ______(time and date)______

About how long will it take you to complete the takeoff? __________

Option 2: We can deliver plans to you ______(time and date)______.

When can you return them?__________

Please advise us of any items that you will not include in your work that we need to include elsewhere.

Figure 2-4. Sample Postcard Invitation to Bid

From______________________________

Contact______________________________

Project______________________________

Bid date______________________________

We will complete our cost estimate ______(time and date)______. This Invitation to Bid requests your quote for ______(description of subcontract category)______

____

____

on the ______________________________ project.

Please call to arrange to use the plans and specifications in our office or pick them up. Because of the number of bidders and sets of plans available, plans can be borrowed for 24-hour periods and weekends only.

Many builders hesitate to use new or relatively inexperienced subcontractors. Other builders prefer to get the service of a small-volume, ambitious trade contractor and exchange assistance with managing the work or material purchases for lower prices or a more prompt response.

Over time, fair subcontracting practices yield the best bids and performance from subcontractors. "Bid shopping" on cost estimates can have short-term benefits but negative long-term effects. Better specialty trade contractors adjust their bids to builders known for these practices. Likewise, slow payment to subcontractors, mismanagement of subcontractors, or scheduling problems on a recurring basis also will raise the subcontractors' subsequent bid amounts.

Many builders like to lock in subcontractors' prices for a number of houses or for a fixed period of time. This procedure is most often used by high-volume builders for labor-only or unit-cost subcontracts when the subcontracted work is similar in size, quality, and scope.

However, builders should be careful in establishing the terms and conditions of such agreements with subcontractors. Builders may not want an agreement that is binding for all the houses that they build. In addition, the specialty trade contractor may want to control the maximum number of subcontracts in force at any time. (The multiple-contract agreement with a specialty trade contractor is a topic beyond the scope of this book. A knowledgeable construction attorney can best advise readers on this topic.)

If a builder and a subcontractor appear "married" to each other, getting other fair subcontract bids may be difficult, or the subcontractor may have difficulty getting opportunities to bid work for other builders. This arrangement may be fine, as long as the contractor's prices remain competitive. However, a lack of competition in subcontract bidding can increase that trade contractor's bid price over time.

Many medium- to large-volume builders spread work around among a number of subcontractors; they obtain multiple bids for each subcontracted trade for each construction contract. Other builders have standing arrangements with two or three subcontractors for each trade and spread the work among them. This arrangement ensures that the builder has another subcontractor to turn to if one should cease to perform satisfactorily or temporarily be too busy to handle a particular job.

Collecting Material Costs

Collecting material costs simply means preparing a complete list of materials that the builder will furnish for the house to be built and determining the cost of each separate item. The quantities and the unit prices comprise the estimate of the total cost of all materials to be installed by the builder's crews or by the subcontractors who do not furnish all their own material.

Most materials are purchased by a standard unit and must be listed as such on the quantity takeoff to complete the cost extensions needed for the cost estimate. Developing a material takeoff worksheet allows quick pricing for the estimate and provides a complete list of items needed by the craftspeople and supervisors who will build the house. The worksheet usually has detailed material information for purchase control.

Because materials are one of the principal items listed on the quantity takeoff, a builder must understand the cost of each type of material to be estimated before beginning the quantity takeoff. The checklist and the sequence of the quantity listing should be arranged for simple, quick, and accurate pricing for the material estimate to be valid. The takeoff must be accurate, for the material estimate to be valid.

To estimate the cost of a particular material needed to build a house, the estimator measures or counts the quantity on the plans and multiplies it by the unit cost of the material.

Some materials are available for purchase on a lump-sum basis with the supplier or vendor preparing the list of material quantities. This type of material cost can be used in the estimate without preparing a list of quantities only if the cost quotation and purchase order or contract states, "all of the material required to complete the requirements per plans and specifications for the house."

Many lumber and building vendors are willing to prepare a list of materials as part of their quote for the material cost of a house. When material will be purchased as a number of units at so many dollars per unit equaling some total cost and the vendor's price is stated that way, the price quotation is actually a unit-cost bid, and it guarantees the unit cost. However, a vendor does not stand by the quantities because he or she has no control of materials after delivery. Therefore, the builder must verify that the quantity estimated by the vendor is satisfactory for the construction project involved.

Figure 2-5 lists the various aspects of material costs, and they are discussed in subsequent paragraphs.

Figure 2-5. Aspects of Material Costs

- Material allowances, including how to calculate allowances for specific material items
- Material specifications
- Calculating material loss, including computations for loss of specific materials
- Lump-sum and unit-cost bids
- Material prices versus quotes
- The material cost summary
- Material purchase control
- Material discounts
- Sales taxes

Using Material Allowances

An allowance is a lump-sum or unit-cost estimate for a material or item that still has to be selected. Any cost overage or savings from the allowance is charged or credited to the home buyer.

In the contract and speculative building markets, beginning the construction of a house before all of the materials have been selected is standard practice. Finish materials such as carpet, hardware, and light fixtures, are likely to be among those not yet selected. To establish a contract price and complete the estimate, a builder must ask the customer what quality of material will be selected. Using an allowance permits the builder to assign increases in cost to the home buyer if he or she chooses something that costs more than the allowance.

Figure 2-6 provides examples of material cost allowances.

Figure 2-6. Material Cost Allowances

Appliances—Allowances are used for optional appliances, such as a whirlpool bath or a garbage compactor, more often than for standard appliances. Although the home buyer may have a choice of color, many builders prefer to control the brand and vendor selected rather than buy appliances with an allowance.

Bricks—Unit cost allowance of $_______ per thousand bricks. Specify the quantity by cost to allow the home buyer to select the color, style, and quality.

Roofing—Unit cost allowance of $_______ per 100 square feet of roofing (may include additional labor for wood shakes or specialty roof). Specify the quantity by cost to allow the home buyer to select the color, style, and quality.

Electrical fixtures—Lump-sum allowance of $_______ a unit. Count the number of fixtures and assume an appropriate cost for each. The allowance is on a lump-sum (total cost) basis to simplify the cost accounting to the home buyer.

Flooring—Carpet unit cost allowance of $_______ per square yard, with allowances for other flooring to be in $_______ per square feet or square yard depending upon how it is sold. Because the quantity of each type of floor is known, you may prefer to use a lump-sum allowance, instead of this unit-cost approach.

Finish Hardware—The allowance for finish hardware is usually a lump sum based on the estimated quantity multiplied by the cost of standard quality. Many builders prefer to specify and provide the hardware, instead of using allowances.

Kitchen cabinets—A lump-sum allowance of $_______ for kitchen cabinets to allow the home buyer to select the type, style, and grade of kitchen cabinets and countertops. The actual cost of the cabinets selected should be compared to the allowance prior to ordering the cabinets and countertops.

Wall Coverings—A lump-sum allowance of $_______ for wall coverings, including wallpaper and special paint to allow the home buyer to make the selections.

Each allowance is based on an actual or assumed quantity takeoff multiplied by the assumed unit cost. To develop the assumed quantities or unit costs the estimator should contact suppliers and vendors or refer to unit costs on recently completed houses of the same size and quality. A sample computation of an allowance for electrical fixtures appears in Figure 2-7.

If the unit costs in Figure 2-7 are accurate, and the home buyer decides to use an overhead fan with a light instead of just a light in the master bedroom, the builder will charge the home buyer an extra $80.

In such situations, builders charge for overruns or give credit for underruns and furnish receipts for material (and labor on combined allowances) to be compared to stated allowances. Comparisons are on a unit-cost basis for unit-cost allowances or a lump-sum basis for lump-sum allowances.

Combined labor and material allowances are used in circumstances

Figure 2-7. Sample Computation of Allowance for Electrical Fixtures

Location	Fixture	Number	Cost/Unit	Total
Living Room	Fan/light	1	$100	$100
Dining Room	Light	1	$40	$40
Kitchen	Light	1	$30	$30
Bath 1, 2	Heat, Vent. Fan,	2	$45	$90
	and Light	2	$15	$30
Bedroom 1, 2, 3	Light	3	$20	$60
				$350

in which the cost of the labor will change depending on the material selected. For instance, the labor cost to install a wood shake roof is more than that required to install an asphalt shingle roof. The allowance should mention whether labor cost is or is not included to prevent any misunderstanding when the actual cost is completed.

Cost allowances in the estimate and contract require extra effort during construction. Therefore, if a home builder cannot expedite the choices, he or she should limit the opportunities for home buyer choices because any delay in selection can cause a construction delay. The contract should specify that, if all selections are not made by a certain date, the home buyer will incur additional costs resulting from the delay.

Many builders control this process by restricting customers' stock choices. The kitchen cabinets and countertops might be selected from a single, limited-selection catalog, and electrical fixtures and flooring, from the showroom stock of a single supplier. When a sales contract includes other than stock selections, the builder may want to adjust the jobsite cost or margin to include the additional management costs associated with custom selections.

Allowances should be realistic for the quality of the house. A large, luxurious house will usually have higher-quality finishes than a basic, speculative house. Therefore, the allowances for fixtures, flooring, wall covering, and even brick or roofing are higher for such a house.

Some builders use allowances in their marketing plans. When the estimated cost of a house must be reduced, often the allowances are the easiest items to cut. Although low allowances are not always prudent, as stated earlier, some builders purposely use low allowances to market houses on a preconstruction basis. Home buyers are rarely happy when they exceed the allowance amount. The best policy is to provide a reasonable allowance for each item and use it without any marketing adjustments.

In many cases the square-foot or component estimate is based on cost allowances for material quality and, in some circumstances, material quantity. The estimator must then revise, adjust, and refine the estimates as the plans and specifications are completed. Any allowance overruns or underruns will alter the proposed contract amount. If material selections or construction details are still incomplete, some items may remain as lump-sum or unit-cost allowances when construction begins.

Using Material Specifications

Building materials are available in different grades, quality, finishes, and colors. The specification for a material is the principal factor in determining its unit cost.

The takeoff listing for each material item includes a type or quality reference for accurate cost estimating. The information on the quality of materials is found most often in the specifications.

In requesting quotations or prices from material vendors, the estimator should stipulate the quality, as well as the quantity, of material required. The accuracy of the quote or price depends on the accuracy of the information furnished, so the estimator must be certain that the vendor or supplier knows and understands the material specification.

The same is true when soliciting labor and material subcontract quotations.

If a specification is unavailable, the estimator must indicate the specification of the material for the supplier or examine the qualifications of the supplier's price or quotation regarding material quality.

Quality requirements for different materials are expressed in different ways depending upon the material. Lumber is classified according to grade; concrete, according to strength; and asphalt shingles, according to their weight.

Calculating Material Waste or Loss

Before estimating the cost of waste or loss of materials, the estimator must understand the conditions that affect waste and loss. The best guide to anticipated loss is the record of actual loss experienced on previous construction projects involving similar work.

Rules of thumb are dangerous because they may be applied without regard to the conditions, supervision, or labor factors that affect the actual amount of material required to build an individual house. Many builders are aware that certain crews or supervisors need more materials to build a house than others. Because of the methods and processes used, one crew and supervisor may have more material waste than another, but the material waste may be offset by higher worker productivity and a lower total cost.

For items that can be counted on the plans, the quantity takeoff should be exact, and no waste or loss factor is required. Doors, windows, cabinets, shelves, rods, accessories, and hardware that can be counted from the floor plan will be lost only through theft or accident. The loss of even a single unit can be costly. Controlling the supply and installation of these items reduces the chance for theft.

Methods of Calculating Quantities for Materials to Be Modified

Some materials are purchased in a particular size that must be modified to fit the house plan. The modification often causes loss or waste. The quantity takeoff listing must account for these losses. Methods of calculating quantities for materials to be modified are discussed below.

Lumber—Lumber is bought in 2-foot lengths; therefore, each takeoff item should be rounded off to the next highest increment of 2 feet to account for waste. For example, a span of 15 feet 4 inches requires a 16-foot piece. The amount of waste changes with design conditions, and therefore, it cannot be accurately calculated as a simple percentage.

Sheet Material—Plywood, particle board, sheet flooring, sheetrock, siding, and other materials are available in certain size sheets. The size of the sheet must be compared to the area to be covered to determine the anticipated waste. The estimate would specify the number of sheets or pieces required, rather than the square footage.

Subfloor and Roof Sheathing—Before calculating the area to be covered, the estimator should increase the length to a multiple of 4 feet and the width to a multiple of 2 feet before calculating the area to be covered. The estimator then divides the calculated area by the area of one sheet, and rounds up the total to the next whole number to determine the number of sheets.

Example—If the subfloor for a house is 26 feet 4 inches by 44 feet 8 inches, and 4 × 8-foot sheet material is used, the calculation is as follows:

$$28' \times 48' = 1{,}344 \text{ square feet}$$

$$\frac{1{,}344 \text{ sq. ft.}}{4' \times 8' = 32 \text{ sq. ft. per sheet}} = 42 \text{ sheets}$$

$$\begin{array}{rl} 1{,}344 & \text{subflooring needed} \\ -\underline{1{,}176} & \text{subflooring area} \\ 168 & \text{sq. ft. of waste} \end{array}$$

Because the actual floor area is 1,176 square feet, the waste included in the quantity take-off is 168 square feet or 14 percent (168 ÷ 1,176 = 14%). For roof sheathing, the best way is to determine the layout required to cover the roof and the overhangs and then count the individual sheets required.

Wallboard—Anyone who has been on a job just after the drywall hangers have finished can attest to the amount of waste. The best way to calculate how much drywall a particular house will need is to calculate the total wall area, including covering up all windows and doors, and subtract any openings wider than 6 feet. Add to this figure the ceiling area, which is usually determined from the floor area plus or minus any unusual features. Note any other special conditions such as fire-rated or moisture-resistant drywall. Special hung areas such as furr downs and other architectural details must be taken off separately.

Concrete—Cast-in-place concrete quantities are seldom less and often more than quantities determined from the dimensions on the plan. To determine expected versus excessive waste or loss, estimators must know the yield of the concrete. (Consult Table 1 in Appendix A.)

Most concrete for houses will be poured onto graded earth or perhaps granular fill. Some inconsistency in the bottom surface that produces variances should be expected, but converting this variance to a strict percentage is difficult.

Example—A 1-inch variance on a 12-inch-deep footing is 8 percent, while a 1-inch variance on a 4-inch slab is 25 percent.

Earthen side forms for footings will be inexact. If rain occurs between the time that the excavation is done and the concrete is placed, the wet material must be mucked out, the earthen form gets larger, and it requires more concrete.

Conservative residential estimators add 5 percent to the volume of concrete slabs and flatwork and 10 percent to concrete footings. Some estimators figure the loss as 1 cubic yard per pour and use the exact quantity calculation plus 1 cubic yard for each planned pour.

Lump-Sum Bids

Lump-sum material quotes are preferable to unit costs for estimating purposes. Materials bid on a lump-sum basis require an accurate quantity takeoff, but none is required for materials purchased on a lump-sum basis. The vendor submits a single cost bid for all materials required for a category or item of work, such as trusses or cabinets.

In choosing which bids to include in their estimates, builders want bids that are easy to compare because they need to verify that all items are included. Therefore, all vendors submitting lump-sum bids must have complete and identical information.

The initial sorting of items for vendor bids should separate the unit-cost items from the lump-sum items. The request for these bids should stipulate, "Bids are to be in lump-sum form to furnish all material required to complete the ________[item of work]________ for the ____[description of residence]____." Many estimators prefer to list the lump-sum material items and quotes on the cost estimate sheet separately from unit-cost materials.

Unit-Cost Bids

Most materials are purchased on a unit-cost basis, which requires the completion of a quantity takeoff for each item. Builders must be certain that the quotes and prices received are in the same units of measurement to simplify comparison. This uniformity eliminates the need to convert or adjust the unit costs.

To accurately compare the material bids prior to entering them on worksheets or cost estimate sheets requires a separate list and file for collecting the unit costs of the various materials.

Builders must keep the costs of items such as doors and windows that are obtained from vendor catalog or price lists separate from other estimated material costs. Because of the time limits imposed by these sources, these costs are material quotes rather than material prices.

Some material categories, such as concrete and gypsum board, are subject to fluctuations in price, so they are usually bid on a project-by-project basis. You should note any qualifications of a proposal. Often these qualifications are standard.

Example—Ready-mix concrete companies often allow 10 minutes per cubic yard for truck time at the site, and they add an additional amount if the actual time exceeds that allowance.

If the vendor furnishes bid prices rather than quantity quotes, the estimator must consider using material cost contingencies to allow for any cost escalation. Price bids differ from quotations in that the estimator may use a guaranteed quote instead of a price bid to protect the cost estimate from increasing, even if the quote is slightly higher than the price.

Some vendors, such as suppliers of framing lumber and wallboard, complete their own takeoffs to support their material unit price or quote bids. Their bids are unit-cost proposals, not lump-sum. You should compare the vendor's quantity listing to your own quantity determination before using such a bid.

Material Prices Versus Quotes

A material quotation is an agreement guaranteeing delivery of material for a specific cost at a designated time. The agreement may limit the time until delivery (often to 90 days).

A material price is the cost of an item at that point in time. The delivery price is subject to any interim price changes.

During an inflationary period or on lengthy building projects, material quotation agreements are necessary to reduce the possibility of cost overruns. The quotation may be on a unit-cost basis, such as "$48.65

per cubic yard of 2,500 psi concrete with no additives." Or the quotation may be a lump-sum "to furnish all of the doors, windows, and finished hardware for $2,721" for a house. In most cases the quote is valid only if the builder accepts it within 30 days from the date of quotation. When a quotation is accepted, it becomes a contract between the vendor and the builder, and the builder may be obligated to buy the material at the cost set forth in the quotation. If prices decrease or the builder finds better prices elsewhere, he or she may be unable to take advantage of them. Therefore, reading the terms and conditions of any agreement is crucial prior to accepting a material cost quotation.

When a builder elects to use current material prices instead of requiring a cost quotation, he or she is anticipating that prices will not increase before the materials are bought for the construction of the house. The builder risks a cost overrun if a price actually increases, but also benefits from, any price reductions on materials.

You must be sure that each material item estimated can be bought for the quoted cost. If quotes are used, the cost estimate will be accurate. When only material prices are available, the estimator may want to anticipate price increases and adjust the cost estimate. Accurately forecasting price increases is difficult and risky.

Builders must examine the price or quote for a total package of materials, such as framing or doors and windows, to determine if it is a lump-sum or unit-cost quotation. Many builders have used more materials than were listed in a package quote and then learned that it was actually a unit-cost agreement and ended up paying for the difference in quantity.

When builders depend on information from others to estimate the cost of the materials for houses, builders must understand all of the information provided before they can complete a valid cost estimate.

Material Cost Summary

The purpose of preparing a total material cost summary is to review and validate material costs on a broad basis, prior to closing out the cost estimate. The summary of material costs also helps in creating the estimate breakdown for cost control. Many of the material quotes must be accepted within a specific period (often 30 days), with a delivery date scheduled to protect the vendor from cost escalation. When builders can use material cost estimates on their own, they can separate the purchase order process and expedite the purchase of materials.

The simplest way to summarize material costs by categories is to separate the work items on worksheets or cost estimate sheets by type of material or work.

Material Purchase Control

Builders must use every effort to purchase materials within the cost budgeted. Material purchased in excess of the estimated unit-cost will result in a cost overrun.

The material purchase control system should be based on the material cost estimate. (See Chapter 7, "Computerized Estimating," for details.) Builders can achieve overall material cost control by purchasing every item for the house according to the estimate.

Figure 2-8 lists ways to prevent material cost overruns.

Figure 2-8. Ways to Prevent Material Cost Overruns

- Include all materials required for the house in the cost estimate.
- Figure a reasonable amount for waste and loss when estimating material quantities.
- Purchase only the estimated quantity.
- Use accurate quotes and prices, do not guess.
- Make estimate cost extensions error free.
- Purchase the material from the low-bid vendor at the amount quoted or priced.

Material Discounts

Many suppliers and vendors offer a discount (usually 2 percent) on materials when invoices are paid within a certain period of time, usually 10 days from the invoice date. So, from an estimating and management viewpoint, builders potentially have two different costs for the material. Prompt payment means lower actual costs.

Many builders price their estimates at full cost and keep the benefit of any discounts as additional profit margin for their companies. In this case, if the quantity takeoff is correct, the estimate includes the maximum cost for the material, and eliminates cost overages.

If builders figure their material estimates at a discount, they must buy that material according to the terms and conditions of the discount. When payment is not in accordance with discount terms and conditions, the discount cannot be taken, and a cost overage will occur because the estimate was based on discounted prices. If a builder has a contract with a cost-plus agreement, the home buyer pays the actual cost and receives any benefits from discounts for prompt payment. However, in competitive bid situations, builders may need to bid with discounted prices or quotes. If you have a history of missing discount dates, using the list price is usually better than using the discounted price. But if you normally make the discount dates, the discounted price is a more accurate estimate of what the cost will actually be.

Sales Taxes

All materials purchased for construction of a house are subject to local and state sales taxes. The prices or quotations furnished by the supplier or vendor do not usually include sales taxes, especially for unit-cost prices and quotes.

The prices and quotes in direct-cost estimates also usually exclude sales taxes. To avoid multiple calculations, most estimators figure the taxes based on the total material cost. Requests for prices or quotations must state whether sales tax is to be included in the price or quote. Also, when the price or quote is received, the estimator must verify whether or not sales taxes have been included.

All subcontracts involving material should state that materials and applicable taxes are the responsibility of the subcontractor furnishing the material.

Labor Cost Estimate

Many builders do not estimate the cost of labor for the houses they build because they subcontract all of the craft work to be done. However, they benefit from understanding how a subcontractor estimates labor costs. Builders can easily monitor the progress of the subcontracted work and keep track of crew size to learn a subcontractor's actual labor cost.

Example—A masonry subcontractor provides a crew that costs $30 an hour in raw labor costs. This crew builds an 11,000-unit brick-veneer house in 1 week. The subcontract for labor is $175 per 1,000 bricks. The subcontractor's raw labor cost is 40 crew hours x $30 = $1,200. The subcontractor must also pay labor burden (Social Security, Unemployment Compensation tax, Worker's Compensation Insurance, and benefits). If the labor burden is 15 percent, the total cost to the subcontractor is $1,380. Because 11,000 bricks at $175 per 1,000 = $1,925, the masonry subcontractor's markup is $545 or 28 percent of the contract.

The same process is applied in reverse to convert hourly labor costs into unit costs for bidding work items when the labor is paid on an hourly basis. As an alternative, builders might buy individual labor on a piece-work basis with a unit-cost labor subcontract.

Labor cost measurements are based on payroll records and knowledge of work completed. Recording labor in workhours at the jobsite provides the initial information used for paying workers. With 1-week pay periods, builders can easily compare payroll costs with work completed during the same week to determine the unit cost of work. However, they must recognize the potential inaccuracy in cost recording and use the information judiciously for any special or unusual circumstances.

Managing labor costs involves comparing measurements with the labor cost estimate. Coding the labor cost estimate, as well as payroll information on the project, according to the NAHB Chart of Accounts allows the builder to compare anticipated costs to actual costs. (Those builders who are unfamiliar with this system, can consult *Accounting and Financial Management* published by Home Builder Press, National Association of Home Builders.[3]) The benefits of this system include timely cost information for productivity and cost evaluation, as well as accurate information to use in future estimates of labor cost.

The following paragraphs discuss preparing the labor cost estimate, including unit cost of labor, labor productivity, labor allowances, wage rates, summarizing and evaluating labor cost, overtime, and labor burden.

Unit Cost of Labor

A worker or crew costs a specific amount of money for a period of time, such as an hour, a week, or a month. On the average, that worker or crew will complete a certain amount of work in that time. The estimator uses the average amount of work performed in a given time period to calculate the labor cost on a unit-cost basis.

Example—A crew that is paid $1,000 a week and usually completes 250 units of work per week has a work item unit-cost of $4 a unit ($1,000 ÷ 250 = $4).

A builder usually thinks of labor on a workhour or crewhour basis, but a unit-cost method is preferable for estimating labor cost. Therefore, the builder should consider developing a system for talking about and evaluating labor cost on a unit-cost basis. The quantity takeoff usually lists work items according to the quantity of materials required. Therefore, the builder has to convert hourly labor costs into costs per unit as listed on the quantity takeoff to simplify completion of an accurate cost estimate.

Example—Sheetrock (gypsum board) is measured by square footage of wall area and bought by the number of sheets required to complete the work. If a crew costs $32 an hour and completes the installation of 320 4 × 12-foot sheets (15,360 square feet) of sheetrock ceilings and walls in 6 days (48 hours), the unit-cost is figured as follows:

$$4' \times 12' = 48 \text{ sq. ft} \times 320 \text{ sheets} = 15{,}360 \text{ sq. ft.}$$
$$\$32 \text{ an hr.} \times 48 \text{ hrs.} = \$1{,}536$$
$$\frac{15{,}360 \text{ sq. ft.}}{\$1{,}536} = \$0.10 \text{ a sq. ft.}$$

The principal formula involves workhours or crewhours, hourly wage, and productivity measured in material units.

To develop labor unit costs that are accurate for use in the estimate, record the labor costs and the productivity in crewhour units. (Chapter 6, "The Cost Control System," covers the measurements and calculations used in the cost control system.)

When thinking about construction, a builder figures that a $32-an-hour masonry crew lays 10,000 bricks in 1 week; when estimating, he or she should consider that paying $1,280 to lay 10,000 bricks is $128 per 1,000 bricks.

Usually estimates should include the cost of installation of all materials under labor cost. Estimators often omit the labor cost for materials purchased on a lump-sum basis. This labor may be estimated on a unit-cost basis and require a quantity takeoff, even though the materials are purchased on a lump-sum basis.

Productivity of Labor Crews

The actual cost of labor for a work item depends on the productivity of the crew. The unit cost estimate is the result of multiplying the wages paid to workers on an hourly or daily basis by the amount of work (in units) expected to be accomplished in that period of time.

The cost estimate becomes the budget and the productivity requirements for the work. Because different crews and supervisors have different productivity levels, unit costs will differ. The weather also affects productivity for certain activities. However, the estimate and the budget must include a reasonable number of crewhours or workhours to complete the work, and the workers must complete their tasks in that amount of time or less to avoid a cost overrun on that work item.

Example—Most automobile repair shops use standard, published labor estimates to calculate repair costs. The repair estimate gives the total of the standard number of workhours allowed for each task multiplied by the wage rates charged by the shop. Therefore, most repair estimates are similar and often exactly the same for the same wage rate. Mechanics who are more productive than the standard productivity rate make more money for their shops than expected. On the other hand, shops with less-productive workers make less and may even lose money.

If the unit cost estimates for labor involve paying crews and supervisors on an hourly basis, you must measure and record the productivity of each worker. The national references for residential estimating standards use average productivity rates of crews. These standards provide guidelines for work items with which a builder has no experience. (A discussion of these references appears in "Published Cost References," in Chapter 5, "Accuracy in Estimating"; also see "Selected Bibliography" at the end of this book.)

To control the actual labor cost in the field, builders must control workers' productivity by using the estimates as budgets of crewtime. Many builders close out their labor cost estimates by comparing the total labor cost to the average crew cost. They use the resulting number of crewhours, -days, or -weeks to estimate labor costs.

Example—The total labor cost for a house is $19,200, and a builder's average crew makes $60 an hour, $480 a day, or $2,400 a week). The budget for labor is 320 crewhours or 8 crewweeks:

$$\$19{,}200 \div \$60 = \$320 \div 40 = 8$$

Labor Allowances

Labor allowances are similar to and are often combined with material allowances. You may use material and labor allowances on a unit-cost basis for preliminary negotiations on first-cut or component cost estimates. These approximate values are appropriate for the level of detail of the plans and specifications available.

Computing job costs on a gross basis for allowances is easy, especially for unit-cost subcontractors. Often, you only need to divide the entire cost of the item by a single-unit measurement. However, you may prefer to combine, rather than separate, labor and material allowances and use a combined unit cost for the estimate. Typical examples appear in Figure 2-9.

Figure 2-9. Examples of Materials and Products with Labor and Material Costs

Material or Product	Material Costs and Labor
Flooring	$/sq. yd. (installed)
Painting	(interior) $/building sq. ft. floor area (exterior) $/siding sq. ft. area
Roofing	$/square (installed)
Hardware	$/door (installed)
Millwork	$/1 ft. (installed, finished)
Light fixtures	$/each
Framing labor	$/sq. ft. of total area
Framing material	$/sq. ft. of total area
Masonry	$/1,000 bricks

Using large-scale allowances on a unit basis involves the quantity takeoff and cost-extension method of estimating. This process is more consistent and accurate than some other forms of estimating.

As with material allowances, discussed earlier in this chapter, the cost to the home buyer increases if the amount of work units or bid-unit costs increase. Using allowances in a first-cut estimate provides for systematic revision of the estimate through completion of the plans and specifications, and an accurate cost estimate results.

Wage Rates

Because you know and record the wage rates paid to supervisors and workers, you can anticipate increases over a short period of time. Fortunately, residential building projects usually take only a few months to complete, so builders usually do not have to forecast increases in wage rates. Most builders use current wage and productivity rates for estimating labor cost on houses to be built. This approach requires that productivity increases match any wage increases to maintain the estimated unit cost for labor.

An increase in wages increases the hourly cost of the crew. This change creates a similar increase in unit costs, unless productivity increases. Many builders use the same workers year round and have had essentially the same crews for years. The wages paid are usually negotiated on an individual basis, rather than by collective bargaining. To stay competitive, builders usually approach wage increases conservatively.

Some builders give bonuses to superintendents based on savings on the labor estimate. An underrun on labor cost indicates higher than anticipated productivity, unless the actual wage rates are less than anticipated, or the contingency cost estimate is too high.

Labor Cost Summary and Evaluation

The total of the labor costs for all work items becomes the labor budget for building the house.

Figure 2-10 provides a sample bid summary for labor cost. To determine the total labor cost estimate for the house, a builder lists the total labor costs from each worksheet or cost estimate page, and adds them.

You must evaluate the labor cost total because it is a principal cost risk. One evaluation method compares the total estimated cost to the total labor cost incurred on similar houses that have been completed. The comparison allows the builder to understand the estimate of labor costs on a complete project basis. Builders can also divide the total labor cost by the average crew cost on an hourly, daily, or weekly basis to understand the time budgeted for the work.

Labor Overtime

The cost of overtime may be justified on certain homebuilding work items, particularly work that is likely to be damaged by rain or cold temperatures. Using overtime in anticipation of bad weather is probably less expensive than using it after such weather impedes the building process. In such a case, a builder may elect to use overtime on some critical activities.

Figure 2-10. Sample Bid Summary

BID SUMMARY
JOB: KINGSTON RESIDENCE JOB No. 162

ESTIMATE SUMMARY	LABOR	MATERIAL	EQUIPMENT	SUB-CONTRACT		TOTAL
PAGE 1	$ 1876 -					
PAGE 2	2955 -					
PAGE 3	14066 -					
PAGE 4	- 0 -					
SUBTOTAL	$18,897 -					
LABOR BURDEN 26% FICA, WORKMAN COMP. UNEMPLOYMENT	4911 -					
SUBTOTAL	$23,808 -					

NOTE:
LABOR BURDEN IS PAID ON ALL EMPLOYEES PAID ON AN HOURLY, RATHER THAN CONTRACT (UNIT-COST), BASIS.

Example—A crew earns $60 an hour and installs 100 units per day. The straight-time unit cost is $60 an hour × 8 hours = $480 a day; and 100 units per day = $4.80 a unit. If the crew works 10 hours a day, 6 days a week with a 10 percent reduction in productivity during overtime hours, the unit cost is as follows:

	Hours Worked	Cost per Hour	Total	Units
Straight Time	40	$60	$2,400	500
Overtime	20	$90	$1,800	225
			$4,200	775

$4,200 ÷ 775 units = $5.42 a unit

The labor cost increases approximately 20 percent when the work week shifts from 40 to 60 hours with a 10 percent decrease in productivity for the extra hours. However, completing the work in fewer calendar days decreases the exposure to bad weather. Material, supply, and equipment costs of the particular work items are unaffected by overtime.

If you plan to pay overtime, you should budget for it in the estimate. Keeping the cost estimate separate helps control the actual overtime cost. For convenience, many builders estimate the unit cost and apply a percentage factor for overtime. (See the preceding example.)

Labor Burden

The estimate of labor must account for all of the indirect costs of labor burden, including Social Security, Worker's Compensation Insurance, Unemployment Compensation tax, and benefits. The standard labor burden ranges from 20 to 30 percent of direct labor costs with the largest variable being the worker's compensation rate.

The wage rates used in unit-cost estimates of direct labor usually are based on the wages paid to workers. However, a builder pays others additional costs for workers, and they must choose a method that accurately forecasts these additional costs and include them in the estimate.

Many builders prefer to use a percentage of the total direct labor cost for figuring the labor burden in the estimate. This method allows the estimator to total all of the direct labor cost estimates and then calculate the labor burden once for the whole job.

Social Security—Labor burden includes matching employees' contributions to Social Security. The rate changes from time to time, and such changes must be reflected in the labor-burden calculations. The Social Security contribution at the time of this writing is 15.3 percent of wages paid, and the employer's portion is 7.65 percent.

Workers' Compensation Insurance—Rates for Workers' Compensation Insurance rates vary according to the craft, a building firm's experience rating, and the locality in which the business operates. Because builders employ a variety of craftspeople, they may apply the exact rate to each craft, or they may develop and use an average Worker's Compensation rate. To appropriately adjust the Worker's Compensation component of the estimate, builders can compare previous compensation rates to current rates.

Unemployment Compensation—Computing an Unemployment Compensation tax multiplier is similar to computing Worker's Compensation Insurance. Builders should review current rates and forecast increases to determine the estimated rate.

After calculating the total direct labor estimate on the cost summary, you should figure the estimated cost of the labor burden as a percentage of the direct labor cost. (The labor estimate summary shown in Figure 2-10 includes a calculation of the labor burden.)

Tools, Supplies, and Equipment

The cost estimate for equipment, tools, and supplies is often based on the time of use or factors other than the amount of work to be done. The format and approach to completing this part of the cost estimate must be accurate and complete.

The first and most important step is to develop a complete list of equipment, tools, and supplies that the builder furnishes to support the work items, possibly including some subcontracted work items (as opposed to those furnished by the subcontractors or workers). When builders provide the same tools, supplies, and equipment consistently, the list is the same for each house.

These costs, estimated from the cost records, depend upon careful recordkeeping for accuracy. For this category, estimators sometimes separate the worksheet and the cost estimate form. Cost estimates based on time depend on the plans, the schedule, and the amount of work required. Therefore, builders often estimate this category of work after completion of the quantity takeoff and the labor cost estimates.

Tools

A builder who employs labor on an hourly basis often furnishes some of the tools needed for building. Some builders combine the cost of supplying tools with equipment costs when estimating. Estimating tool costs separately provides more accurate and comprehensive costs than the combined method.

Cost provides one basis for defining or categorizing tools for estimating purposes.

Example—Items that cost less than $250 may be classified as tools, and those that cost more than $250 may be classified as equipment. This approach may or may not suit a particular builder's operation. Because most builders use and replace a consistent inventory of tools, listing items by work type or craft may be easier. Such a listing might include the following items:

Utility Tools—Shovels, post hole diggers, brooms, wheelbarrows, hoses, electrical extension cords

Carpentry Tools—Circular saws (blades, sharpening costs), nail guns (air hoses, compressors)

Concrete Tools—Rakes, hoes, vibrators, finishing equipment (power trowels with blades)

Masonry Tools—Wheelbarrows, shovels, scaffolding (Scaffolding is often included in the equipment inventory with mortar mixers and charged to the job on a rental basis.)

Because tools are replaced and inventory increased as needed, their cost per house is inconsistent. Therefore, the best approach is to estimate the cost on an average basis. Many builders refer to these items on the estimate as a small tool average or allowance. Because such an average must be as accurate as possible, many builders divide their annual tool cost by their annual labor cost to set up the estimating percentage.

Example—A builder's 1991 total payroll cost was $166,500, and total tool cost was $3,330. Therefore, the total tool cost was 2 percent of the labor estimate. Therefore, on a 1992 house with a labor estimate of $8,800, the small tools estimate would be $176.

With this method, the cost-accounting system establishes a tool account. When tools are purchased, they are charged to the account rather than to the job. The system then charges each house for tools used at a standard percentage rate. If the rate is accurate, the builder will recover expenditures for tools throughout the year. However, unless tool costs are listed as a line item in the cost estimate, they are likely to be paid for from the estimate markup.

Supplies

Supplies include items that may not be specifically shown on the plans and specifications but are required to build the house. Supply items are expendable; they are used up on the job.

Building supplies are most often included in the material estimate, the small tool estimate, or the jobsite cost estimate. Builders should make sure that their estimate checklists include expendable items in the categories that will produce the best estimate and the best system of accounting.

Some items are more difficult to assign than others. For example, cleaning supplies, as well as the labor for cleaning, are required to complete the construction of a house. Unless cleaning is a labor and material subcontract, the builder must provide cleaning materials and labor. (The checklist in this book lists labor and supplies for cleaning under jobsite costs.)

Equipment and Rented Tools

To estimate the cost of equipment and rented tools, multiply the amount of time they are rented by their rental rates. Most builders prepare the cost estimate for equipment on a job basis, rather than on an item basis. Equipment and tool needs are determined before doing the takeoff and the labor estimate.

The rental period is figured in hours, half-days, days, weeks, or months. The builder pays for the time he or she has possession of the tool or equipment rather than for the time it is used. Therefore, builders must calculate the cost by anticipating the total time that the equipment will be on the job.

The rental rate is higher on short-term rentals, but the total cost from such rentals can be lower. Therefore, in anticipating length of use, a builder should select the appropriate rental rate.

Example—If a power trowel rents for $45 a half-day, $65 a day, or $160 a week and the job requires 2 half-day pours and 2 full-day pours, the builder should consider the sequence of construction to estimate the cost. If a substantial amount of time is to elapse between pours, the calculation would be as follows: 2 × $45 = $90 and 2 × $65 = $130, for a total cost of $220. If all pours can be completed in 1 week, the estimated rental cost would be $160. (Some estimators prefer to use the higher estimated rate with any money saved going to the builder.) Most builders tend to rent the same equipment and tools again and again. In this case the builder can develop a checklist of equipment rental items for estimating.

For accuracy the builder must know whether the rental rate includes or excludes nonhourly costs, such as delivery and pick-up, sales taxes, fuel, insurance. A valid equipment rental estimate qualifies each unit cost.

Owned Equipment

Builders use different methods of accounting for and charging for their own equipment. Most builders "rent" the equipment to the job by the day, week, hour, or month at a predetermined rate. The basic idea is that the home buyer pays for the equipment, rather than the builder.

The best way to keep competitive on the cost of renting equipment is to check rental rates at local rental services and compare them to national published standards.

For accuracy, the total costs for the items discussed below must be converted to an hourly or daily rate for incorporation into the this cost category as part of the hourly or daily "rental" charge.

Replacement Cost of Equipment—For many years contractors used original cost or book cost or value information in determining hourly rates. In theory, a contractor recovers the original cost by the time the equipment is worn out. Unfortunately, the cost of replacement equipment is usually higher than the cost of the original equipment. To estimate a reasonable replacement cost for the purchase of new equipment a builder has to figure out its rental rate.

Insurance—Builders must recover the cost of insurance for their equipment from the income from houses sold. After they figure the total annual cost of insuring the individual pieces of equipment, they should estimate the anticipated annual hours or days of use (based on historical records) and divide that figure into the total insurance cost to determine the amount to add to the "rental" rate for insuring the equipment.

Taxes and Licenses—To figure out the cost of licenses and taxes for "rental" purposes, builders must divide the annual figure by the estimated number of hours or days that the equipment will be used.

Repair and Maintenance—Although anticipating an exact amount for maintenance and repair is difficult, builders must establish a budget for this expense and convert it for inclusion in the hourly or daily charge. The cost of repair typically increases with the age of equipment. Therefore, the maintenance and repair costs for a particular piece of equipment are likely to be higher each year it ages. Some publications, such as guides for earth-moving equipment, have average hourly maintenance and repair costs.

Operating Expenses—Fuel, oil, lubricants, and tires are the principal components of annual operating expenses that must be translated into a daily or hourly rate for inclusion in the equipment rental budget.

Support—Builders must account for the storage and transportation costs of equipment in charging expenses to projects. This cost is often established as a lump-sum move-in, set-up, or mobilization cost and charged according to the number of times the equipment is moved to the site, rather than by an hourly or daily rate.

Operator—Many builders include the cost of operating equipment in the labor cost estimate. Other builders prefer to separate these costs and include them in the hourly or daily rate. You can compute these costs once each year (See Figure 2-11.) and use the rental rate established for all estimates and charges to projects for that year. A yearly review of the income and expenses for each piece of equipment should validate the equipment charge rates. Likewise, if you pay cash for equipment, the loss of interest income that the cash would generate must be figured in.

Some items, such as fuel, may not be measured separately for each piece of equipment. Builders often combine the budgets for these items into a single accounting category, which they can measure and evaluate separately.

To forecast income and expenses for equipment for the next fiscal year, you should compare the previous year's figures for the anticipated and for the actual number of hours or days used. To validate the charges on an individual basis, you can divide the equipment expense by actual use (in hours or days) for each piece of equipment and compare it with the rate charge for that item.

If you borrow money to purchase equipment, finance charges can be substantial. Therefore, you must include the cost of financing in the established charge rates. Likewise, if you pay cash for equipment, you must figure in the loss of interest income that the cash would generate.

Renting Versus Owning Equipment

Builders make diverse decisions about equipment management, depending on their business needs. At one extreme are those who own no equipment and require subcontractors to furnish the equipment needed for their individual work. These builders rent equipment for material handling and work items that are not subcontracted.

At the other extreme are builders who own all the equipment needed for construction except for specialized subcontractor's equipment. When a subcontractor uses a builder's equipment and this use was not included in the estimate and cannot be charged to the home buyer, the subcontractor should pay the builder for using it.

Most builders own only the equipment they use on a regular basis and rent other equipment when they need it. The most economical practice is to own a limited number of each piece of equipment. Unless additional rental equipment is available, however, work on one project may be slowed while workers wait for equipment that is being used for another task or project.

Basically builders must take two considerations into account when trying to decide whether or not to buy a piece of equipment. First, if they are using a particular piece of equipment regularly, paying more for rental than the equipment costs makes no sense. Therefore, if

Figure 2-11. Sample Equipment Rental Rate

EQUIPMENT RENTAL RATE

CASE 580 BACKHOE	1	2	3 ANNUAL COST	4 RATE	5 BREAKEVEN HOURS	6 1040 HOURS PROFIT/LOSS
REPLACEMENT COST						
USED	$10,000 -					
SALVAGE	4000 -					
4 YR. LIFE	6000 -		$ 1500 -			
INSURANCE			700 -			
TAXES/LICENSES			160 -			
REPAIR/MAINTENANCE			850 -			
FUEL/OIL			300 -			
SUPPORT			250 -			
ANNUAL EXPENSE			$ 3760 -			
MARKET RATE				$ 25/HR	155 HRS	$ 22,180 -
BID RATE				$ 15/HR	258 HRS	$ 11740 -

builders see that rent will match or exceed the cost of equipment in the foreseeable future, they should consider buying.

Maintenance is the second factor in the rent-versus-buy decision. Maintaining some equipment such as bulldozers, and backhoes can be expensive if a builder does not have in-house expertise or know a mechanic who charges reasonable prices.

The primary criterion for equipment planning, however, is the amount of time and the frequency with which the builder uses a piece of equipment.

Example—One workyear provides 2,080 potential workhours (straight time). Some builders own only the equipment they plan to use for more than a set amount of time, such as 1,200 hours per year. However, buying a second or third piece of the same equipment makes sense only when other pieces of the equipment are fully used.

A builder typically owns the equipment listed in Figure 2-12.

Figure 2-12. Equipment Builders Typically Own

- Dump/flat-bed trucks (for hauling)
- Utility tractors of various sizes, such as front-end loaders, box blades, scrapers, and backhoes
- Fork lifts, high-lift material handlers, man hoists
- Scaffolding
- Trenching equipment

Equipment Cost Summary

Builders can more accurately review and account for estimates of equipment costs when these costs are segregated. Figure 2-13 provides a sample equipment cost summary. In this example, equipment rental costs are segregated from charges for owned equipment to facilitate cost accounting procedures.

Some builders prefer to include the equipment, tool, and supply costs for each item with either the labor or material costs. This practice often adds either labor burden costs or sales tax costs to the cost of rental equipment and reduces the accuracy of the cost estimate.

Some estimating formats provide three cost columns for each takeoff item: equipment, material, and labor. A principal problem with this approach is that the units used to estimate material and labor costs do not apply directly to equipment. Costs for equipment expressed in material or labor units are not accurate for all circumstances and can reduce the accuracy of the cost estimate.

Jobsite Costs

A contract to build a house includes many cost items that are not direct material, labor, subcontract, or equipment costs. These costs are for items such as supervision, temporary utilities and facilities, and builder's risk insurance.

Builders estimate the cost of each of these items, just as they do with material and labor costs.

Figure 2-13. Sample Equipment Cost Summary

EQUIPMENT COST SUMMARY
JOB: SPEC HOUSE #56

EQUIPMENT	HOURS	RATE	TOTAL EQUIPMENT
BACKHOE	8	$ 25/HR	$ 200 –
TRACK LOADER	4	$ 35/HR	140 –
BULLDOZER	8	$ 45/HR	360 –
BOBCAT	4	$ 25/HR	100 –
SCAFFOLDING	20 DAY	$ 20/DAY	400 –
MOBILIZATION 4 PIECES		$100/EA	400 –
EQUIPMENT COST			$ 1,600 –

The total construction time, or a part of the time, determines the cost of many indirect cost items that must be included in the estimate, including those discussed immediately below:

Supervision—The superintendent is paid on a weekly or salary basis, including benefits. To figure the cost of supervision, divide the superintendent's annual salary by the number of houses he or she builds in a year.

Temporary Utilities—Previous job records provide the basis for estimating the cost of telephone, power (or generators), water, and temporary toilets on a monthly basis. Builders may assign some mobilization, hook-up, or set-up charges and costs to subcontractors, but usually they must include some of them in the estimate.

Temporary Facilities—Different projects require different support facilities for construction. A checklist for these items would include a guard or watchdog; office and storage trailer or buildings, sheds, or other job-built facilities; temporary fencing; and tree protection. Some builders estimate these costs on a lump-sum basis. Many others prefer to do it on a monthly basis and use previous job records for the estimated cost.

Layout—To estimate the cost of layout, multiply the labor cost by the anticipated time required and add the cost of supplies, material, and equipment required.

Building Permits—Fee schedules and regulations for permits and utility hook-ups can be obtained from the appropriate inspection and enforcement agencies.

Bonds—Some contracts require performance, completion, and payment bonds. The cost of the bonds is based on the total contract amount. Firm quotes will be needed for the cost estimate.

Insurance—The allocation of costs depends on the method of buying insurance. Builder's risk and other insurance is bought on a project-by-project basis, and builder's risk and public liability insurance costs are included in the jobsite overhead. General business insurance bought on an annual basis must be divided by the number of houses built each year and charged as a jobsite cost. (However, some builders account for this item in markup instead.)

Builders often include numerous other types of costs (such as sales commissions and closing costs) in jobsite costs, rather than as estimated unit-cost items. Builders often estimate clean-up costs on a lump-sum basis because they are difficult to estimate on a line-item basis.

Figure 2-14 is a comprehensive listing of items that may be included in jobsite overhead. An architecturally designed house often has "general condition" and "supplemental condition" requirements that involve costs to the builder that are not included in line-item cost estimates. Therefore, builders develop the general condition cost estimate on a separate cost estimating sheet.

Markup

Markup refers to the percentage or the amount of money included in the estimate in addition to the direct cost of subcontracts, material, labor, equipment, and jobsite overhead. The markup is divided into

Figure 2-14. Sample Jobsite Overhead Estimate

JOB-SITE OVERHEAD ESTIMATE — PAGE 1 OF 2

	ITEMS	ESTIMATE	ACTUAL	VARIANCE
1	ADMINISTRATIVE			
2	PERMIT			
3	INSPECTIONS			
4	WATER FEE			
5	SEWER FEE			
6	POWER			
7	TELEPHONE			
8	TEMPORARY TOILETS			
9	WATER			
10	INTEREST			
11	PLANS			
12	INSURANCE			
13	LIABILITY			
14	COMPLETED OPERATIONS			
15	BUILDER'S RISK			
16	WORKER'S COMP.			
17	(IN LABOR BURDEN)			
18				
19	JOB SITE			
20	SUPERVISION			
21	LAYOUT			
22	SURVEY			
23	SMALL TOOLS			
24	WATCHMAN/PATROL			
25	EROSION CONTROL/SILT FENCE			
26	CLEAN UP			
27				
28				
29				
30	SUBTOTAL PG. 1			
31				
32				
33				
34				
35				
36				
37				
38				
39				
40				

	ITEMS	ESTIMATED COST	ACTUAL COST	VARIANCE
1	CONSTRUCTION LOAN			
2	APPRAISAL			
3	LOAN ORIGINATION FEE			
4	TITLE EXAM			
5	TITLE BINDER			
6	DEED PREPARATION			
7	DOCUMENT PREPARATION			
8	CLOSING FEE			
9	INTEREST			
10	(IN ADMINISTRATIVE)			
11				
12	LOT CLOSING			
13	RECORD DEED			
14	TITLE EXAM			
15	CLOSING FEES			
16	PROPERTY TAX			
17	LAND COST			
18				
19	SUBTOTAL			
20				
21	SALES CLOSING			
22	APPRAISAL (2ND)			
23	PHOTOS			
24	CLOSING FEE			
25	DOCUMENT PREPARATION			
26	NOTICE OF COMPLETION			
27	LOAN ORIGINATION FEE			
28				
29	SALES COMMISSION			
30				
31	ADVERTISING			
32				
33				
34	PAGE TOTAL			
35				
36				
37				
38				
39				
40				

two parts—general overhead and profit. General overhead is also commonly called home-office overhead. That term refers to a builder's headquarters office in contrast to a field office. It does not refer to an office in a home unless, of course, the builder's headquarters is in his or her home.

General overhead includes all expenses not charged to individual projects such as executive salaries, salaries of other home-office personnel, company vehicles, home-office rent or depreciation, telephone, and office supplies and equipment. Profit is the amount of money a company makes after all the expenses are paid, both jobsite expenses and home-office expenses.

The proper way to figure general overhead is to determine the total of all the home-office expenses for a typical year (a) by looking at the accounting records for the past year, (b) estimating the costs based on current expenses, or (c) using a combination of the two. After you determine the total general overhead expenses, estimate your dollar volume for the year and the amount of work you need to do during the next year to make that sales volume. Divide the general overhead expenses by the volume to determine the percent of your volume that will go to general overhead. In other words, a certain percentage of each job pays for running the home office.

If you include all the general overhead expenses (especially the executive salaries), the percentage should be somewhere between 4 and 8 percent. If a builder's general overhead is greater than 10 percent, he or she needs to do more work without increasing fixed expenses. Builders who can keep their general overhead to less than 5 percent have a much better chance of being successful.

An acceptable amount of profit depends on market conditions. The profit added to the cost estimate requires judgment and understanding of the total business. Therefore, the principal decision-maker most frequently determines what the profit will be. The amount of profit should keep a builder competitive without having a big spread between his or her bid and the next one or "leaving money on the table."

If you complete a certain number of houses each year at a relatively fixed mark-up, you will realize more profit by controlling administrative costs. The cost estimate should help to control costs during construction of the house.

Your cost accounting system must provide an annual summary of the cost of administering construction and of administering your business so you can allocate it to your building projects. (Administrative costs include those associated with land, finance, and sales, which are beyond the scope of this book.)

Some builders prefer to assign the administrative costs equally to projects by the number of projects started within that year. Others are most comfortable distributing the cost of administration on a monthly basis for construction of each house. But most builders assign the administrative costs as a percentage of the direct cost, using the total annual value of construction.

Example—If a builder can successfully market 15 houses a year that cost $120,000 with a 15 percent markup that includes both profit and general overhead, the sales volume is $1,800,000. If annual general and administrative costs are $90,000, the general overhead is 5 percent

(90,000 ÷ 1,800,000). Therefore, the profit is 10 percent because 5 percent of the markup is taken up by general overhead.

In most bid situations, the cost of administration is included in the markup or margin, rather than calculated as a direct cost. To summarize:

Subcontract cost summary
Material cost estimate
Equipment cost estimate
+ Job site overhead
Direct construction cost

Direct construction cost
Contingency construction cost
+ Markup or margin
Estimate total or bid proposal amount

Markup
− Cost of administration
Potential profit

Potential profit
− Cost overruns above contingency
+ Cost underruns
Profit

In most markets, the amount of markup on a marketable house is limited. Therefore, the annual profit potential of a company depends on the type and number of houses built.

The following table shows that net annual profit is more a result of the annual value of construction than a result of the number of units built. The variation in the number, size, and quality (expressed in dollars) of the houses to be built is a result of the company's management strategy (Figure 2-15).

Figure 2-15. Effect of Annual Profit on Annual Value of Construction Versus Number of Units Built

Annual Value of Construction	Average House Price	Average Profit	Number of Houses	Annual Profit
$800,000	$40,000	$4,300	20	$88,600
$800,000	$100,000	$11,100	8	$88,800
$800,000	$400,000	$45,000	2	$90,000

Chapter 3

The Quantity Takeoff Process

The quantity takeoff is the process of determining from the working drawings the amount of materials required for the house. This process, detailed in this chapter, is the foundation of the cost estimate. After you have your quantity takeoff and your quotations based on available information, you may then determine the cost of the house. The steps involved in doing that are listed below:

1. Reviewing working drawings.
2. Reviewing the specifications to make sure that prices used match material and work requirements for an accurate cost estimate.
3. Taking measurements of the parts of the house from the working drawings.
4. Checking off items on the plans.
5. Doing the math to determine the quantities for the cost estimate.

The following discussion is intended to familiarize you with the vital estimating tools—the architect's scale, calculator, and checklists. It also describes the steps listed above.

Architect's Scales

You will need to measure quantities of materials from the working drawings, which usually are drawn to scale. Scales used for working drawings vary, depending on what information is presented.

The floor plan, foundation plan, roof framing plan, and elevations are usually drawn at these scales:

$$1/4'' = 1' \text{ or } 1/8'' = 1'$$

Details of foundations, walls, cabinets, and other items requiring greater detail are drawn to a larger scale, usually:

$$1/2'' = 1', 3/4'' = 1', \text{ or } 1\,1/2'' = 1'$$

These scales are referred to as architect's scales, and estimators must use them properly to achieve an accurate quantity takeoff.

All of the scales are multiples of 1/16 inch and ordinarily have two scales on each edge (Figure 3-1). One scale is read from left to right with the division of a single foot on the left end. The other scale is either half or twice the scale of the first, and it is read from right to left with the division of a single foot on the right end.

The estimator must develop an accurate technique in using the scale to maintain accuracy when measuring takeoff amounts in higher quantities. When a scaled dimension does not correspond to a written dimension, the written dimension is considered the correct one. However, the written dimension may not always be correct. An estimator who finds such a mistake, must be careful to determine which is correct.

Many dimensions and combinations of dimensions are used repetitively. Therefore, carefully measuring plans and recording dimensions helps to ensure an efficient and accurate takeoff. A single measurement error can result in several inaccurate quantities and produce an inaccurate estimated cost.

The exterior dimensions shown on the floor plan give the exterior wall length as measured to the outside face of the stud line. This standard practice allows dimensions for exterior wall framing to be derived from the floor plan. Interior walls may have to be scaled because the written dimensions do not give complete information. Scaled dimensions usually are good enough because of the necessary waste factors.

Estimators must be certain to select the correct scale for the drawing to be measured. The scale for the drawing is listed on the sheet, usually below the plan or detail to be measured. The quickest way to check the scale is to compare it with a dimension shown on the drawing. If they are equal, the scale selected is correct.

Figure 3-1 shows measurements made with an architect's scale. The smaller scale is labeled left to right from zero with marks every foot. Only the even-foot marks (shown on 3⁄8-, 1⁄2-, and 1 1⁄2-inch scales) or marks every fourth foot (shown on 1⁄8″ and 3⁄32-inch scales) are noted. One-foot divisions are provided on the left end. To measure, align the right end of a measurement with a marking on the scale, so that the left end of the measurement falls between the zero and the scale designation. This alignment puts the left end of the measurement within the divisions of the foot. The inches are read from left to right.

Figure 3-1. Architect's Scales

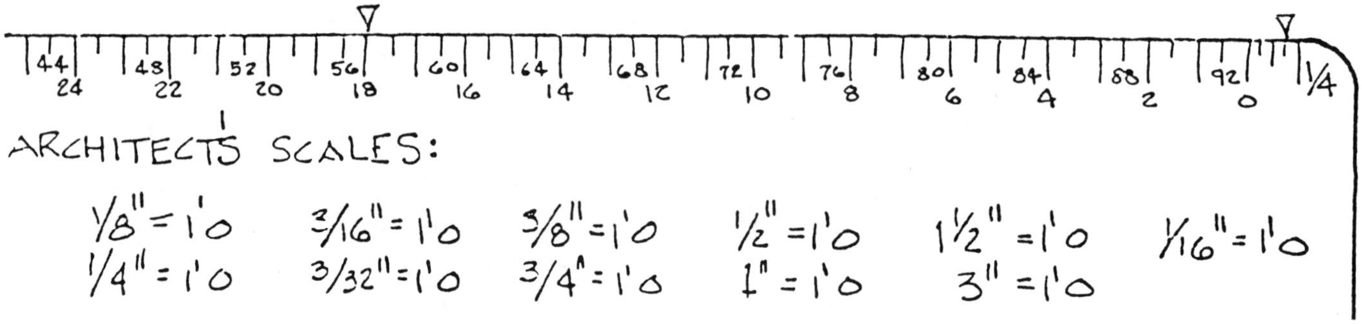

The large scale is labeled from right to left from zero, with marks every half foot, because the smaller-scale foot marks are half as far apart as the foot marks on the larger scale. The left end of the measurement must align with a foot mark, rather than with a half-foot mark. The right end of the measurement aligns to the right of the zero mark and within the 1-foot division.

If the scale matches your drawing, you can use it exactly as a full-scale ruler or tape is used on a house being built. You should use only the correct scale on each plan or detail. If you make all measurements, align the drawing, then read the scale according to the procedure outlined, you should produce an accurate quantity takeoff.

Sometimes the drawings depict some items that are "broken" to fit on the plan. However, a broken dimension must not be scaled. The draftsperson must write the dimension on the detail when such a drawing is used for reference.

You must record the measurements accurately for the quantity takeoff to be accurate. Accuracy also requires rechecking all measurements and comparing them to the dimension written on the plan, especially any measurements that are questioned.

When measuring the length of the walls, some estimators take all measurements drawn horizontally on the plan and then do all of those drawn vertically. They check off each item as they measure it. Other estimators prefer to measure the drawing in a clockwise or counterclockwise sequence and check off each measurement after listing it on the takeoff. Using a standard, repetitive approach on each drawing for each house helps to ensure accuracy and completeness. Properly checking off each item eliminates counting measurements twice.

The beginning estimator can practice using the architect's scale on plans with accurate dimensions. This practice allows the beginner to check measurements. Good scaling habits produce confidence in measuring for quantities of items needed.

Calculator

Most builders are familiar with the functions of a calculator that are needed for determining quantities of materials needed: multiplication, division, addition, and subtraction. Because of the many entries made, this part of the quantity takeoff and cost estimate presents the greatest chance for error.

One approach to reducing errors is to always use a calculator with a paper tape, instead of one with an electronic readout. The paper tape allows an estimator to quickly check entries against the worksheets, takeoff listing, and estimate sheets. Numbers are easy to transpose when listing the calculation results on the quantity takeoff. Checking the paper tape is faster and more accurate than doing each calculation.

Most electronic calculators require decimal equivalents of inches for entries and calculating answers based on feet and inch measurements. An estimator can memorize the following decimal equivalents or post them on the calculator:

1″ = .08′	5″ = .42′	9″ = .75′
2″ = .17′	6″ = .50′	10″ = .83′
3″ = .25′	7″ = .58′	11″ = .92′
4″ = .33′	8″ = .67′	12″ = 1.0′

For standard calculations of quantities, the same calculations are often used repetitively. Standard calculations include the following:

Area of rectangle = length × width
Area of square = length × width
Area of triangle = base × ½height
Area of circle = radius × radius × 3.14

$$\text{Area of } 1/4 \text{ circle} = \frac{\text{radius} \times \text{radius} \times 3.14}{4}$$

Volume of rectangle = area × depth
Volume of tube = area × depth

To convert plan measurements to express quantity determinations in correct pricing units requires the following standard quantity conversion measurements:

1 cu. yd. = 27 cu. ft. (3′ × 3′ × 3′)
1 cu. ft. = .037 cu. yd. (1/27 yd)
1 sq. yd. = 9 sq. ft. (3′ × 3′)
1 acre = 43,560 sq. ft.

Framing lumber is often priced by the board foot, which equals the nominal thickness (in inches) multiplied by the nominal width (in inches) and the length (in feet) and divided by 12. To convert linear feet to board feet, use the following conversion factors:

2″ × 4′ = .67 bd./lin. ft.
2″ × 6′ = 1.00 bd./lin. ft.
2″ × 8′ = 1.33 bd./lin. ft.
2″ × 10′ = 1.67 bd./lin. ft.
2″ × 12′ = 2.00 bd./lin. ft.
1″ × 4′ = .33 bd./lin. ft.
1″ × 6′ = .50 bd./lin. ft.

Checklists

For many estimators, a thorough checklist is their most valuable tool for creating an accurate estimate. A checklist organizes the takeoff. Without a checklist the estimator must rely on memory, which eventually results in embarrassing and often costly omissions.

Two checklists, are provided in this section (Figures 3-2 and 3-3); one is arranged by work classification and one by working drawings. The checklist items are the same for both arrangements. A checklist with pricing units appears in "Step 5, Computations," later in this chapter. You will need to develop your own checklist by adding to or subtracting from the given list. If necessary, rearrange the checklist to suit your approach to the takeoff and the cost control system. Your particular requirements should govern the sequence and the amount of information on the checklist; you need only list the items that you takeoff. (Chapter 2 "The Complete Estimate," covers the use of a checklist for subcontract bid control.)

The work classification arrangement (Figure 3-2) of the quantity takeoff is the most widely used checklist framework. Within this arrangement, the sequence of items can be arranged in a number of ways. Many estimators prefer to list the work in the sequence in which it will occur on the jobsite. This procedure helps to "build" the house

Figure 3-2. Sample Checklist Takeoff Arranged by Work Sequence

ROUGH CARPENTRY

Floors

- Sills
- Beams
- Joists
- Bridging
- Headers at openings, blocking
- Subflooring (building paper)
- Ledgers
- Screeds

Walls

- Plates
- Studs
- Headers
- Gable framing
- Blocking, firestops
- Nailers, girths
- Bracing at corners
- Exterior sheathing
- Posts, columns
- Beams
- Furring
- Backing for trim

Roofs

- Ceiling Joists
- Bridging
- Beams
- Headers at openings
- Rafters
- Ridges
- Hip and valley members
- Trusses
- Leveling beam (strong back)
- Collar beams
- Bracing
- Lookouts
- Dormers
- Purlins
- Roof sheathing
- Backing for trim

Stairways

- Treads
- Risers
- Stringers
- Carriage board
- Nosing
- Disappearing stairways

Wall Board

- Exterior siding
- Exterior Paneling
- Gypsum board (sheetrock)
- Prefinished paneling
- Vinyl coverage gypsum
- Ceiling board

FINISHED CARPENTRY

- Posts
- Beams
- Frieze boards
- Fascia
- Soffit
- Casing (door, window)
- Base and shoe mold
- Trim

Millwork

- Windows (and screens)
- Doors—exterior, interior
- Sidelites
- Screen doors
- Door frames
- Shelving
- Cabinets (unit, detailed)
- Stair and balcony railings
- Closet rods
- Track for sliding doors
- Miscellaneous built-ins

Insulation

- Vapor barriers
- Floor (perimeter)
- Wall
- Ceiling
- Roof
- Weatherstripping

ROOFING

- Felt
- Shingles
- Ridgeroll and edging
- Fasteners
- Flashing
- Built-up roofing

MASONRY

- Face brick
- Common brick
- Fire brick
- Mortar
- Concrete block
- *U* blocl (block fill)
- Wall reinforcing
- Wall ties

PAINTING AND DECORATING

- Sealers and stain
- Primers
- Finish paints—interior, exterior
- Wood floor finishing
- Taping, floating, texturing
- Wallpaper

FLOOR FINISHING

- Tile flooring, wainscot
- Carpet
- Sheet vinyl
- Adhesives
- Vinyl tile
- Wood floors
- Concrete paint

FINISH HARDWARE

- Locksets
- Hinges, doors
- Cabinet hardware
- Shelf brackets
- Bath accessories
- Mirrors

MISCELLANEOUS ITEMS

- Gable louvers
- Soffit vents
- Foundation vents
- Flashing
- Gutters and downspouts
- Thresholds
- Door bumpers
- Roof vents
- Brackets
- Lintels
- Masonry wall ties
- Fireplace damper and/or liner
- Straps, inserts
- Weatherstripping

HARDWARE

- Nails
- Anchor Bolts
- Screws
- Hangers
- Gusset plates (gang nails)
- Roof fasteners

ELECTRICAL/PLUMBING

- Outlets
- Switches
- Fixtures
- Heaters
- Vents
- Cover plates
- Appliances
- Water heater
- Hoods for ranges

FOUNDATION/CONCRETE

- Formwork
- Concrete slab/foundation
- Reinforcing bars
- Mesh (welded wire)
- Walks
- Drives

Figure 3-3. Sample Takeoff Checklist Arranged by Working Drawings

SCHEDULES

- Windows and screens
- Doors (exterior, interior)
- Sidelites
- Screen doors
- Door frames
- Casing (door, window)
- Locksets (specification)
- Hinges

THE WALL SECTIONS

- Plates
- Studs
- Headers
- Blocking, firestops
- Nailers, girths
- Bracing at corners
- Exterior sheathing
- Exterior siding or paneling
- Paneling and vertical siding
- Battens
- Insulation
- Gypsum board (sheetrock)
- Trim
- Taping, floating, texturing
- Prefinished paneling
- Base and shoe mold
- Frieze boards, moldings
- Fascia and mold
- Soffit
- Fascia backing
- Lookouts
- Dormers

THE GABLE WALL SECTION

- Plates
- Studs
- Blocking, firestops
- Nailers
- Siding

THE ROOF FRAMING PLAN

- Trusses
- Ceiling joists
- Beams
- Level beam (strongback)
- Rafters
- Ridges
- Hip and valley members
- Collar beams
- Bracing
- Purlings
- Headers at openings
- Roof sheathing
- Insulation
- Roofing felt
- Shingles
- Furring (plates, studs, strips)
- Ceiling Board
- Roof flashing
- Ridge roll and edging
- Built-up roofing
- Gutters and downspouts

THE FOUNDATION PLAN

- Footings
- Concrete slab/foundation
- Reinforcing bars
- Walks
- Mesh (welded wire)
- Vapor barrier
- Formwork
- Drives

THE FLOOR FRAMING PLAN

- Sills
- Beams
- Ledgers
- Joists
- Bridging
- Headers at openings, blocking
- Subflooring
- Screeds

THE FLOOR PLAN AND ROOM FINISH SCHEDULE

- Finish flooring (carpet, tile, wood)
- Adhesives
- Wainscot
- Shelving brackets
- Light fixtures
- Outlets
- Switches
- Appliances
- Hoods for ranges
- Cover plates
- Heaters
- Vents
- Mirrors
- Door bumpers
- Shelving
- Closet rods
- Posts
- Cabinets (unit, detailed)
- Backsplash
- Cabinet hardware
- Miscellaneous built-ins
- Track for sliding door
- Water heater

MASONRY (Elevations and Wall Sections)

- Face brick
- Common brick
- Fire brick
- Mortar
- Concrete block
- *U* block (block fill)
- Wall reinforcing
- Wall ties

PAINTING AND DECORATING (Room Finish Schedule)

- Sealers and stain
- Primers
- Finish paints (interior, exterior)
- Wood floor finishing
- Taping, floating, texturing
- Wallpaper

HARDWARE

- Nails
- Anchor bolts
- Screws
- Hangers
- Gusset plates (gang nails)
- Roof fasteners

MISCELLANEOUS ITEMS

- Gable louvers
- Soffit vents
- Foundation vents
- Flashing
- Gutters and downspouts
- Thresholds
- Weatherstripping
- Roof vents
- Brackets
- Lintels
- Masonry wall ties
- Fireplace damper and/or liner
- Straps, inserts

STAIRWAYS

- Treads
- Risers
- Stringers
- Carriage boards
- Nosing
- Disappearing stairs
- Stair and balcony railings

on paper when preparing the takeoff. This arrangement groups work performed by the same crew and materials that are from one vendor or related vendors. It also helps to complete the estimate breakdown for the builder's cost control. (See Chapter 7, "Computerized Estimating.")

Some estimators find that a checklist arranged by working drawings saves time (Figure 3-3). A takeoff so arranged is completed on a drawing-by-drawing basis.

The checklist is never really complete. To develop a more complete checklist, you must analyze each drawing, looking for items such as smoke detectors and doorbells that are required but not included on the checklist. Add such items to the checklist when they appear on the plans and the takeoff.

Step 1. Review Working Drawings

The working drawings or plans for the house contain most of the information that will be listed on the quantity takeoff. Read each note and study each detail on the working drawings, record any questions regarding materials, specifications, or methods of construction.

Make note of any missing information that you can find, including drawings or details that are incomplete or left out. This practice helps to make the quantity takeoff as complete and accurate as possible under the circumstances. The list of missing or incomplete items helps you to develop qualifications for the cost estimate.

In some cases, you may prepare an estimate from only a floor plan and some verbal or written description of work not shown on the floor plan. You must visualize or sketch the other drawings and details that would be included in a complete set of plans to qualify your bid or to obtain utmost accuracy. You must also indicate your understanding or assumptions in writing so that no misunderstandings develop. An experienced estimator knows what information should be provided in which segment of the working drawings.

This chapter provides a typical, complete set of drawings including the site plan, foundation plan, floor plan, wall sections, floor framing plan, roof framing plan, elevations, door and window schedule, and room finish schedule (Figures 3-4 to 3-12).

The Site Plan

The site plan (Figure 3-4) provides information on the position of the house on the lot and the location and dimensions of other items that are to be built on the site. Most site plans are drawn with an engineer's scale, not an architect's scale, with 1″ = 20′ being the standard.

The site plan includes the items listed in Figure 3-4. The sidewalk details are usually shown on the site plan, but they also may be elsewhere in the drawings.

The site plan provides information for takeoff of both the first and last work that is required in the building process, including site clearing, excavation for the basement (if necessary), and the finish grading, seeding, and planting of the lot. If the site plan does not indicate both the existing and final grades of the lot, the estimator and excavator will have difficultly determining the amount of cut or fill (excavation) required.

Figure 3-4. Sample Site Plan

Items Usually Included in Site Plan

- Location of building (usually two sides) from the property line
- North orientation of the site
- Dimensions of the building lot
- Location of walks, drives, and patios
- Dimensions of driveways and sidewalks
- Radii of curves
- Location and dimension of steps, terraces, and patios
- Finish floor elevation
- Landscaping requirements including seeding and planting

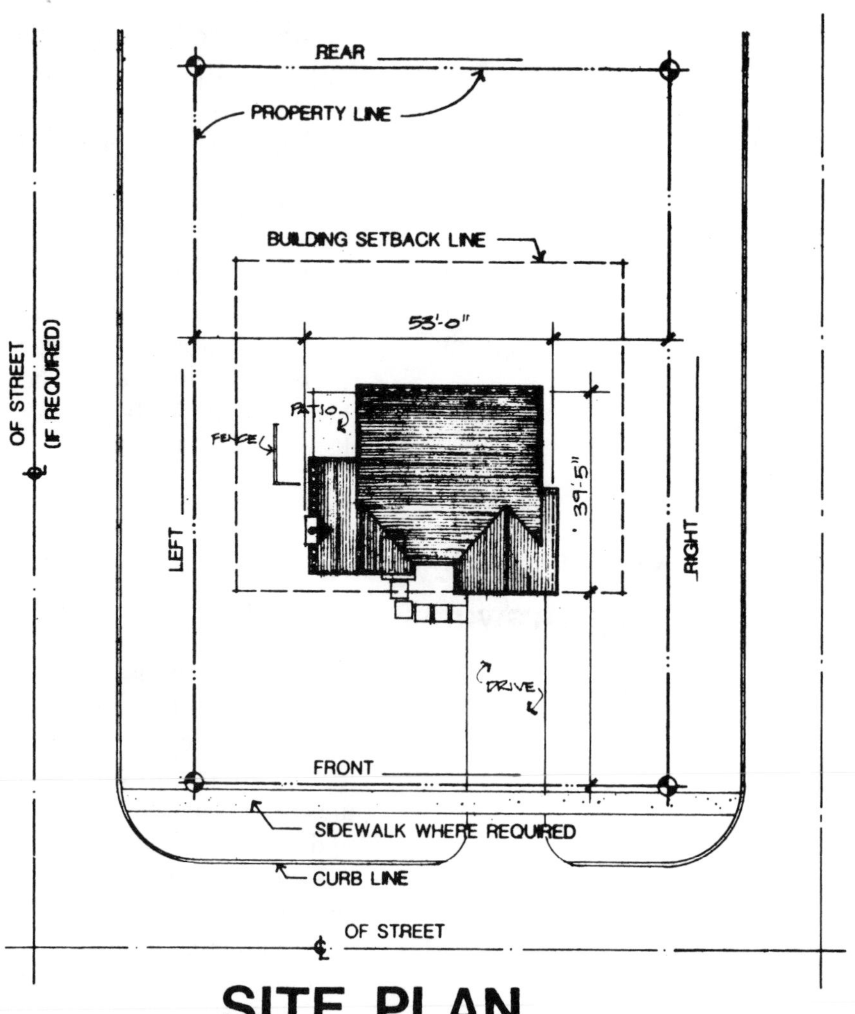

Before starting the takeoff, you should always visit the site to verify that the information shown on the site plan is correct. Without a site visit, you cannot be confident that your estimate is complete. (See Chapter 2 "The Complete Estimate," for more details on the site visit).

The Foundation Plan

The type of work shown on the foundation plan is typically concrete and, in some cases, masonry work (Figure 3-5). Comparing the dimensions on the foundation plan with the dimensions on the floor plan will ensure that the quantities computed for the foundation are correct.

Pay particular attention to details provided, to those left out and to the proper location of each in the foundation. All unusual conditions that affect the cost of completing the foundation work must be noted so that you can adjust quantities and/or unit costs to assure an accurate cost estimate. You should verify that the material requirements on the foundation plan are consistent with those in the appropriate sections of the specifications.

If you subcontract foundation work, review the foundation plan prior to submitting the drawings to subcontractors for their bids. Resolving questions midway through the estimating process is much more difficult than it is before requesting the subcontractor's bids.

The foundation plan is usually drawn to the same scale as the floor plan, typically: ¼″ = 1′. Specific details of the foundation are usually found on the same sheet at a larger scale. The foundation plan includes the items listed in Figure 3-5.

Figure 3-5. Sample Foundation Plan

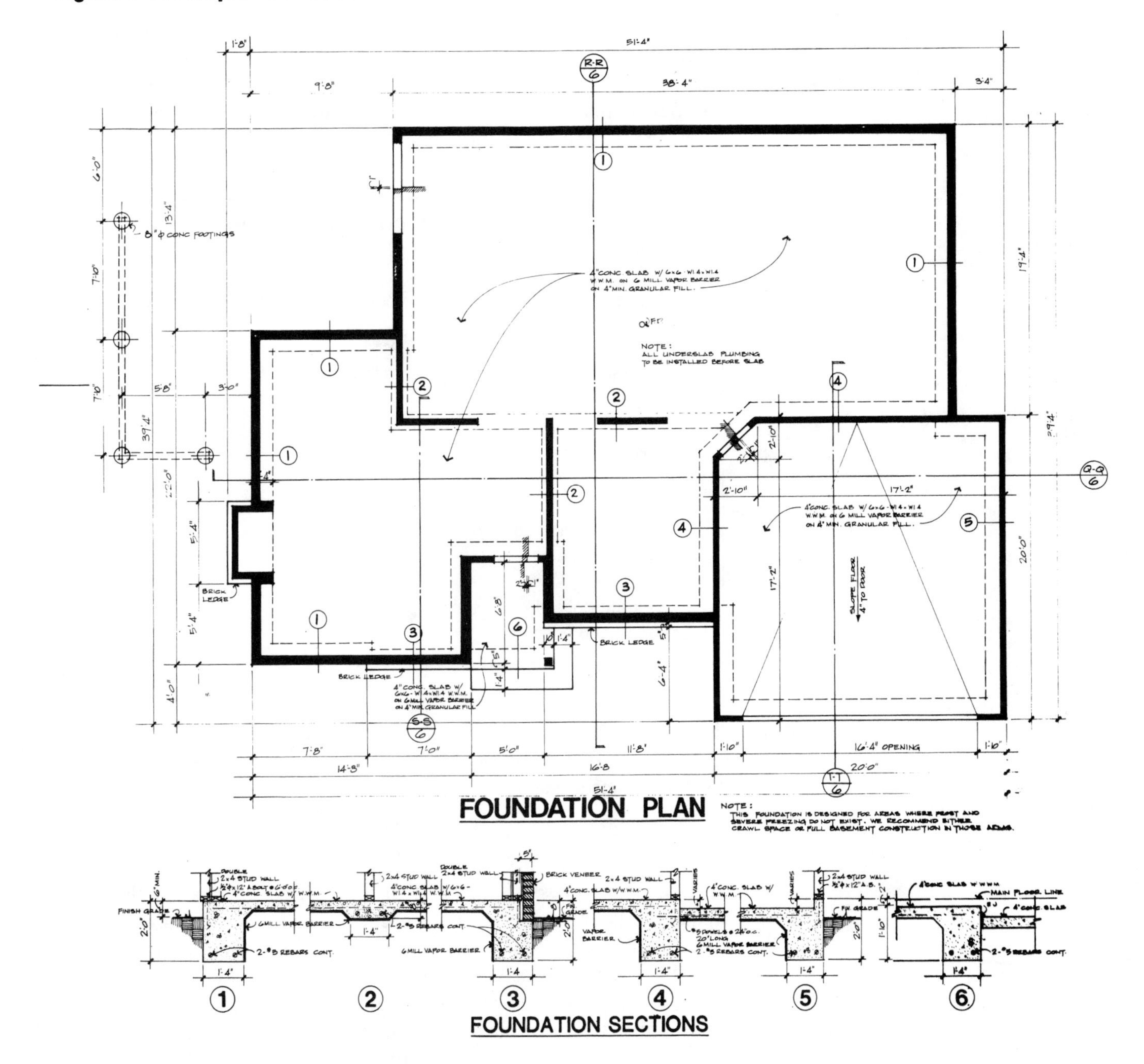

Items in a Typical Foundation Plan

- All necessary dimensions of the foundation
- Footing sizes and locations
- Grade beam sizes and locations
- Foundation wall thicknesses and heights
- Typical notes pertaining to construction; such as 4″ concrete slab, 2,500 psi, with 6 × 6″/10-10 welded wire mesh over 4″ compacted fill or 8″ × 24″ continuous concrete footing, 2,500 psi concrete with 2 #5 bars″
- Exterior slabs for patios and equipment
- Section views (usually to this scale: 1½″ = 1′) of parts of the foundation including typical slab and footing details, thickened portions of slabs under load bearing walls, brick shelves, and notes pertaining to construction.

You should check the dimensions of the foundation plan with the dimensions on the floor plan. If a foundation plan is not provided, you can sketch one using the dimensions from the floor plan and details from other houses of similar construction. The foundation plan might include the items listed in Figure 3-6. Masonry, waterproofing, and insulation details for basement construction, when required, should also appear on the foundation plan.

Estimators need to understand the foundation as a whole and as pieces and parts. The foundation plan is the reference for estimating the cost of the excavation and foundation work required for footings and basements. (Tables 1 and 2 in Appendix A provide quantity conversions for concrete foundations and slabs.)

Figure 3-6. Sample Foundation Plan Dimensions

- Footings
  - Single story; 8″ × 24″ with 2 #5 bars
- Foundation walls
  - Supporting wood frame walls or 8″ block, 8″ wide
  - Supporting wood frame walls or 8″ block with brick or stone veneer, 12″ wide
  - All foundation walls 24″ minimum height
- Thickened portions of slab
  - Under load-bearing walls, 8″ minimum
  - Under columns, 8″ minimum
  - Under fireplaces, 8″ minimum
- Slabs on grade
  - Floor slabs, 4″
  - Garage slabs, 4-6″; thicken edge to 8″
  - Walks and patios, 3-4″

The Floor Plan

The floor plan has measurements for computing the quantities of most materials used in building the house (Figure 3-7). The typical scales for floor plans are as follows:

Typical scale ¼= 1′
Larger houses ⅛″ = 1′

A floor plan shows a house as if it were sliced approximately 6 feet above the floor so that such features as walls, doors, windows, cabinets, shelves are indicated. Because the finish, door, and window schedules are referenced on the floor plan, you may decide to make notes on the schedules while reviewing the floor plan.

The floor plan includes most necessary dimensions including those listed in Figure 3-7, but some dimensions may not be listed on it. Therefore, you should be familiar with standard practices and typical dimensions used in your area including the items listed in Figure 3-7. Many draftspersons include electrical and/or plumbing information on the floor plan and the symbols for these may be confusing. Use the takeoff checklist to help avoid this confusion.

Figure 3-7. Sample Floor Plan

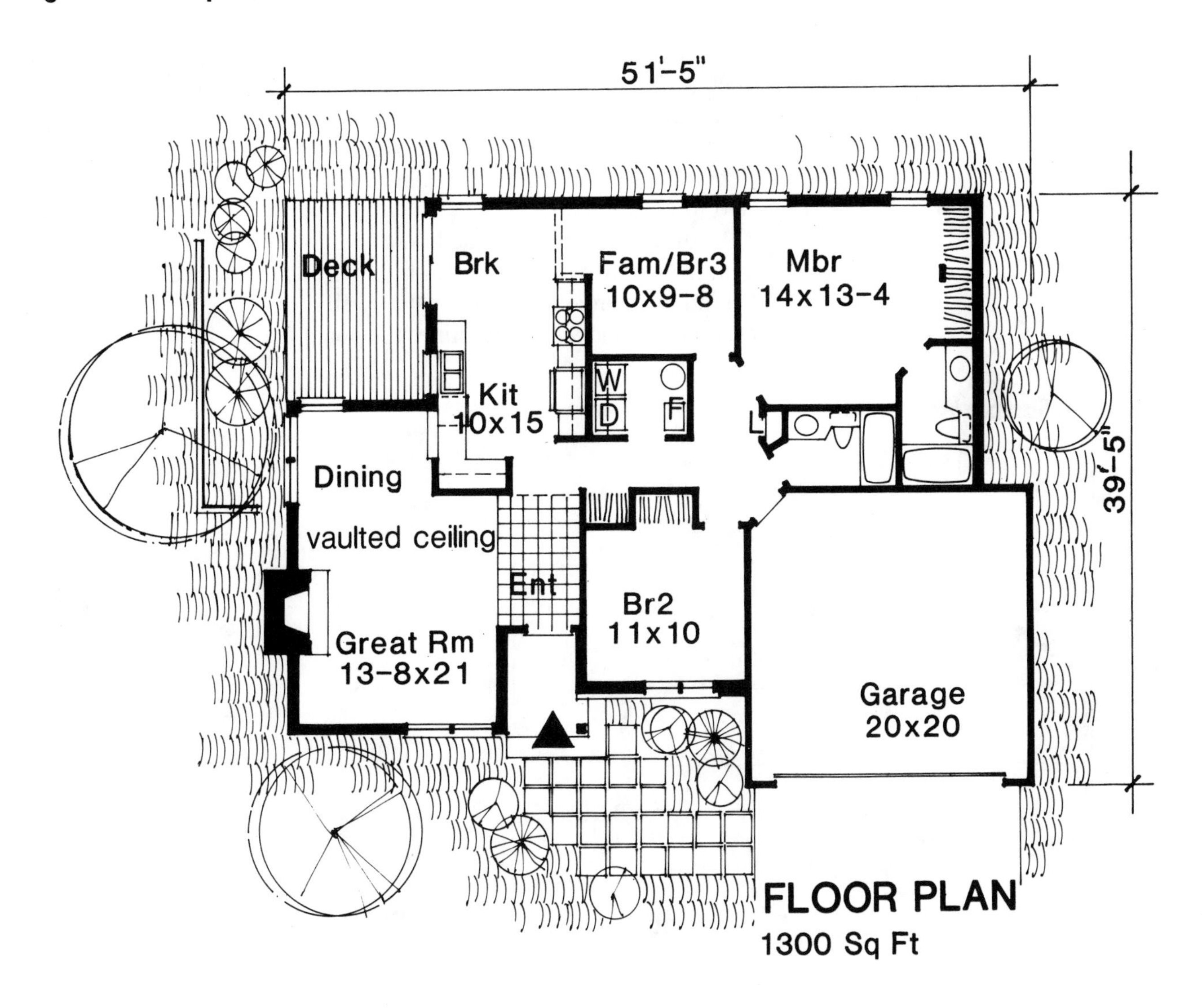

Dimensions on Floor Plan

- Outside wall lengths and widths
- Interior partition lengths, widths, and locations
- Window and door locations (and number)
- Edges of slabs and thickness of masonry wall construction
- Sizes of terraces, walks, and patios (adjacent to the house)
- Special construction items
- Window symbols and identifications
- Door symbols and identifications
- Built-in millwork and cabinets
- Kitchen and bathroom fixtures
- Fireplace symbol(s)
- Special construction features
- Equipment (air-conditioning and ventilation units, heater, and water heater)

Typical Dimensions for Items Not Listed on Floor Plan

- Titles for all rooms/areas
- Concrete block walls
  - Light, 4 to 6″ wide
  - Medium, 8″ wide
  - Heavy, 12″ wide
- Interior wood-frame walls, usually 4-6″ wide
- Brick veneer walls are 4-6″ wood frame plus 2″ of polystyrene insulation and 4″ of brick
- Wet walls, behind plumbing fixtures, 8″ thick
- Closets, usually a minimum of 2½′ deep
- Halls, usually a minimum of 3′ wide

The Wall-Section Drawings

The wall-section drawings (Figure 3-8) provide a clear description of the materials and work required to construct the walls of the house. A different drawing should be provided for each type of wall used, although this practice is not always followed. The wall sections are usually drawn to a larger scale, such as ¾" = 1'. Many times these larger-scale sections are shown broken and cannot be scaled.

The wall-section drawings usually include the necessary dimensions listed in Figure 3-8.

The dimension from finished floor to finished ceiling determines the quantities for many materials. This dimension and the size and spacing of framing members appear on the wall section detail. The cornice detail on exterior walls determines the quantities for framing and back-

Figure 3-8. Sample Wall Section

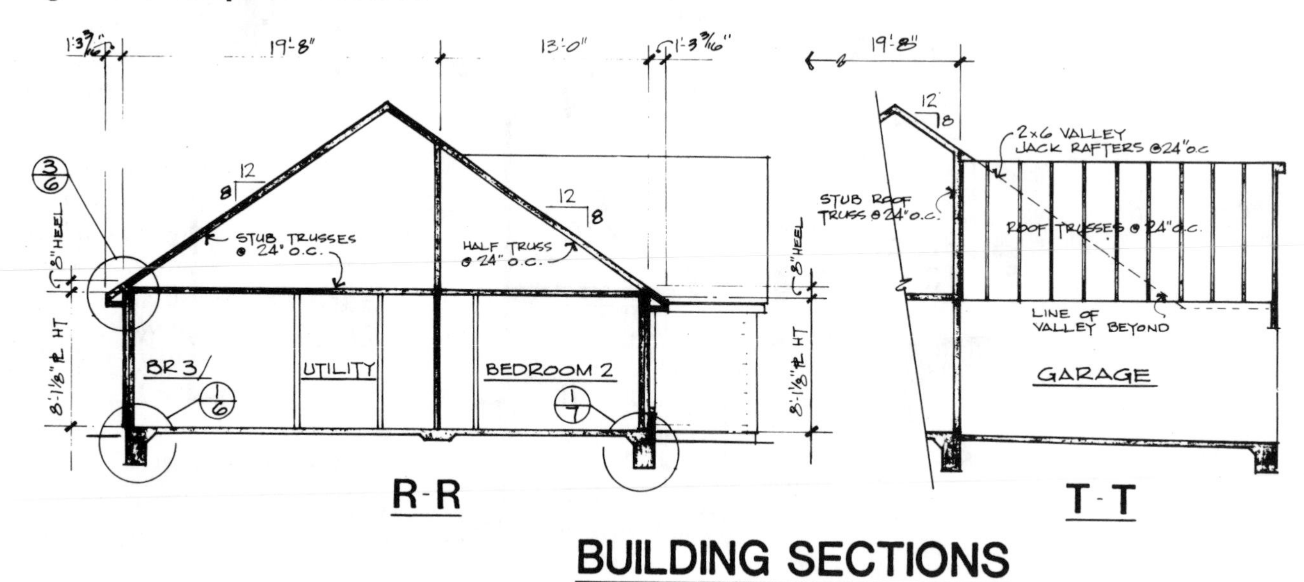

Dimensions on Wall-Section Drawing

Floor to ceiling heights
Depth of floor system (actual or nominal sizes used)
Roof overhang dimensions and materials
Depths of footing(s)

Foundation/Slab

Footings and foundation walls
Floor construction
Compacted fill
Vapor barrier
Reinforcement
Concrete thickness

Sill Materials

Sill plates and anchor bolts
Headers
Subcontract and finish flooring

Wall Construction

Structural material (wood framing or concrete block)
Size and spacing of members
Interior and exterior sheathing and finishes
Siding or brick veneer
Wall insulation
Ceiling materials and finishes
Trim materials
Attic insulation

Cornice details

Gable end framing and siding
Soffit and fascia framing
Roof-framing system
Soffits and fascia materials
Vents
Roof sheathing
Roofing materials

up members, as well as for finish materials required to build the roof, overhang, and porches.

When section drawings of foundations, windows, stairs, or cabinets are provided, they explain the specific materials and methods of construction to be used. You should be familiar with local practices and be able to sketch your own details in case the needed section details are not included with the working drawings.

The Floor-Framing Plan

The floor-framing plan (Figure 3-9) locates, sizes, and notes the spacing of floor joists, beams, headers, sills, bridging, miscellaneous flooring members, and subflooring and flooring area. Each framing member is shown and noted with sills and other details that are unclear elsewhere in the working drawings.

The framing plan provides a picture of exactly which materials and sizes are required. If a floor-framing plan is not included in the drawings, you can sketch your own plan based on the foundation and floor plans.

In using the floor-framing plan to determine the length of the individual pieces required, you should round the length of members needed up to the next 2-foot multiple to ensure that the material takeoff is in the appropriate price units. The number of each type of member equals the number of framing spaces plus one starting piece. Many estimators round up to the next highest 2-foot dimension for the width and the length dimensions of the house or parts of the house before taking off the framing materials.

The Roof-Framing Plan

The roof-framing plan and detail drawings (Figure 3-10) identify the slope, size, location of the roof, and the size and spacing of trusses or joists and rafters, collar braces, ridges, hip members, valley members, leveling beams, and miscellaneous framing and braces.

You will often need details on trusses, collar beams, rafter bracing, dormers, lookouts, and roof overhangs to determine the amount of roof-framing material required. A roof-framing plan included in the working drawings simplifies the takeoff of the framing members.

You should use these roof-framing drawings with judgment because every piece that is needed may not be shown on the plan if an experienced builder should know that the piece belongs there. If a roof-framing plan is not included, sketch the roof-framing layout and use the floor plan to clarify your estimate.

The rafter factors (Table 7 in Appendix A) will help you convert the flat measurement of rafters to the correct length required for the roof. As with other framing member measurements, round these off to the next highest, 2-foot multiple. You can also use the rafter table to convert the flat roof area (multiply length by width) to the sloped area to determine quantities of sheathing, felt, and shingles required.

Figure 3-9. Sample Floor-Framing Plan

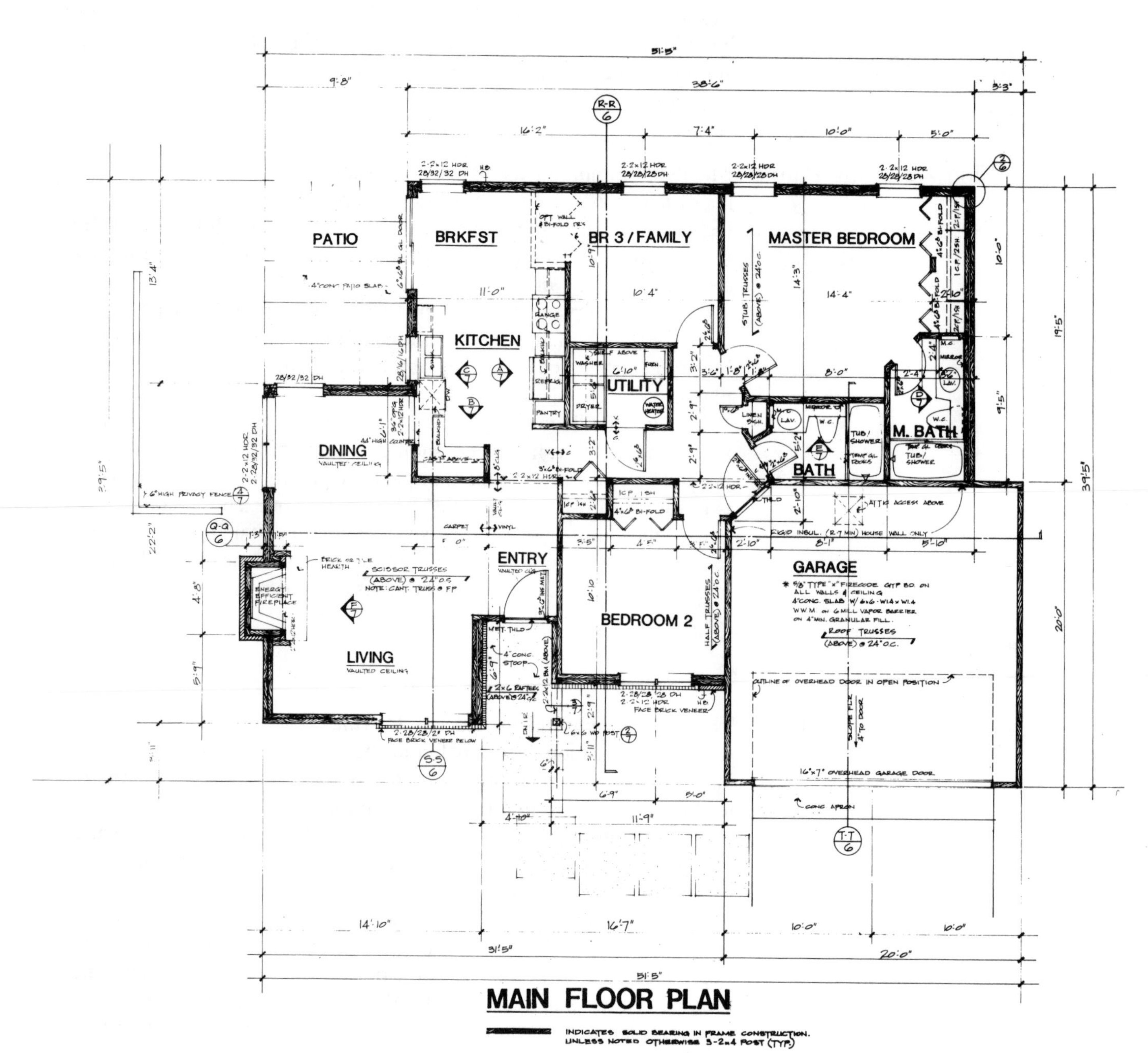

Figure 3-10. Sample Roof-Framing Plan

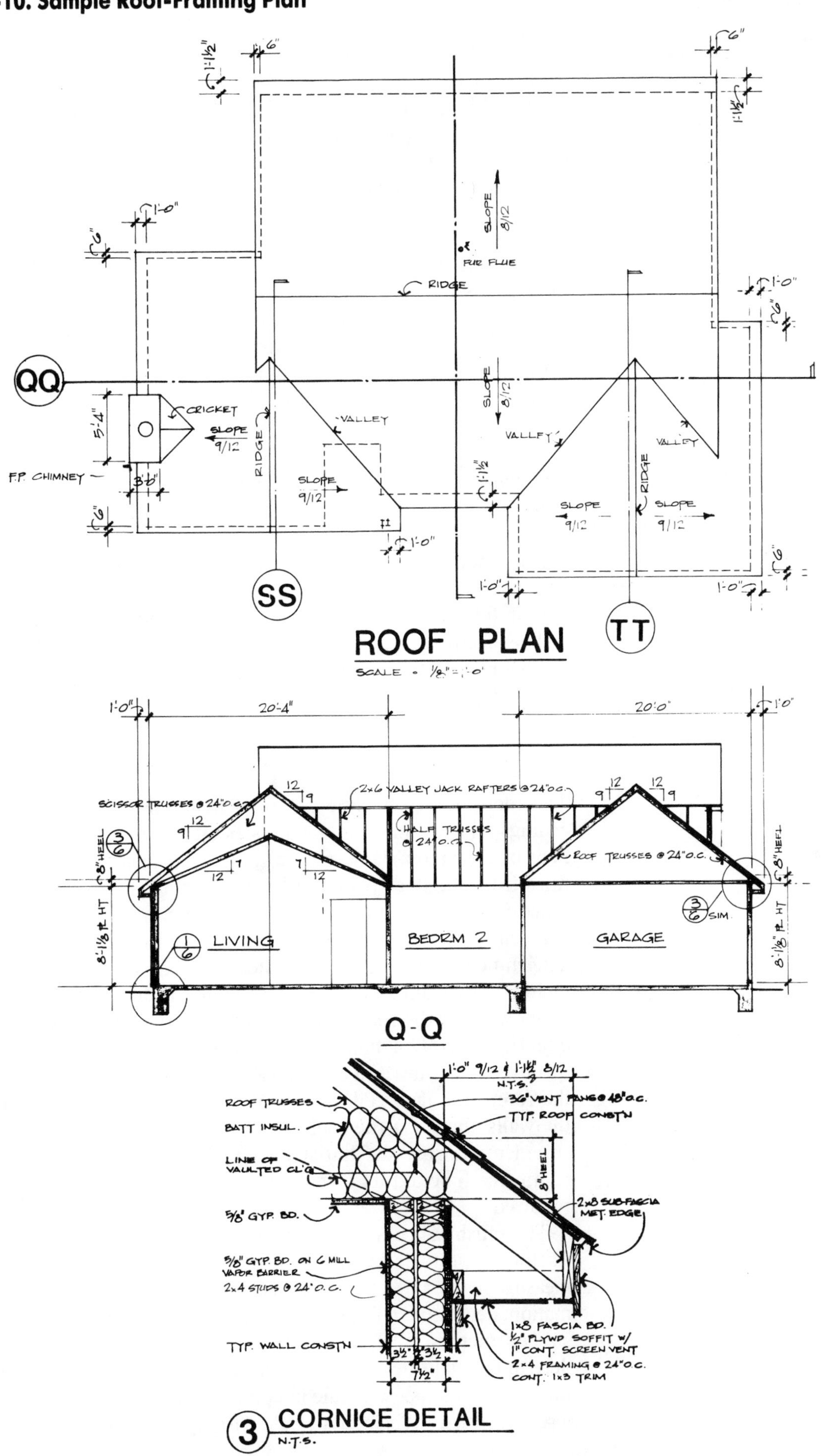

Elevation Drawings

Elevation drawings help estimators, builders, subcontractors, and others to visualize the house to be built (Figure 3-11). Elevations provide a view of the completed exterior of the house and identify the materials used. Before starting the quantity takeoff, you must ensure that dimensions on the wall sections match the dimensions on the elevations. The scale of the elevations is usually as follows:

$$\frac{1}{8}'' = 1' \text{ or } \frac{1}{4}'' = 1'$$

The elevations sometimes but not always include all the necessary dimensions for the items listed in Figure 3-11. If elevations are not included, you should qualify the estimate and specify exactly what you will furnish in the way of exterior materials and details. You can consult previously used plans and specifications to specify finishes and details.

The Window-and-Door Schedule

The schedule of windows and doors specifies exactly what doors and windows go in what openings and describe the requirements for the openings noted on the floor plan and in the elevations. The schedule usually includes information on such items as frames, hardware, finishes, and trim. In some instances, the schedule provides all the information for the door and window items instead of the specifications.

The takeoff includes counting the number of doors and windows on the floor plan and elevations according to type. The quantity takeoff listing of the count should provide a reference to the schedule and specifications for estimating purposes.

The Room-Finish Schedule

The room-finish schedule identifies materials and finishes for walls, ceilings, floors, and trim. This schedule contains references to the floor plan by room name and/or number. You should determine quantities from the floor plan and list the materials and finishes from the schedule on the quantity takeoff, according to the checklist. The floor plan, along with wall-section drawings and other details, is used to determine the quantity of the materials needed in finishing the house, including gypsum and other board materials for the ceiling and wall.

A review of the room-finish schedule and a comparison of it to the quantity takeoff of finish items verifies the quantities needed. The quantities for most finish items are based on either the perimeter of the room, the floor/ceiling area of the room, or the linear footage of the walls enclosing the room. You should check and list these measurements during the initial plan review and compare them to the room finish schedule to calculate the quantities of each material required.

The finish subcontractors, including painting and flooring subcontractors, base their estimates on the room finish schedule, with their measurements and computations from the floor plan, and the information determined from the details. If you use unit or labor-only subcontractors (Chapter 2, "The Complete Estimate"), you must determine quantities needed for finish material and/or work items from the floor and finish schedule in appropriate pricing units and allow for loss and waste. (Chapter 4, "Preparing a Complete Estimate," and

Figure 3-11. Sample Elevations

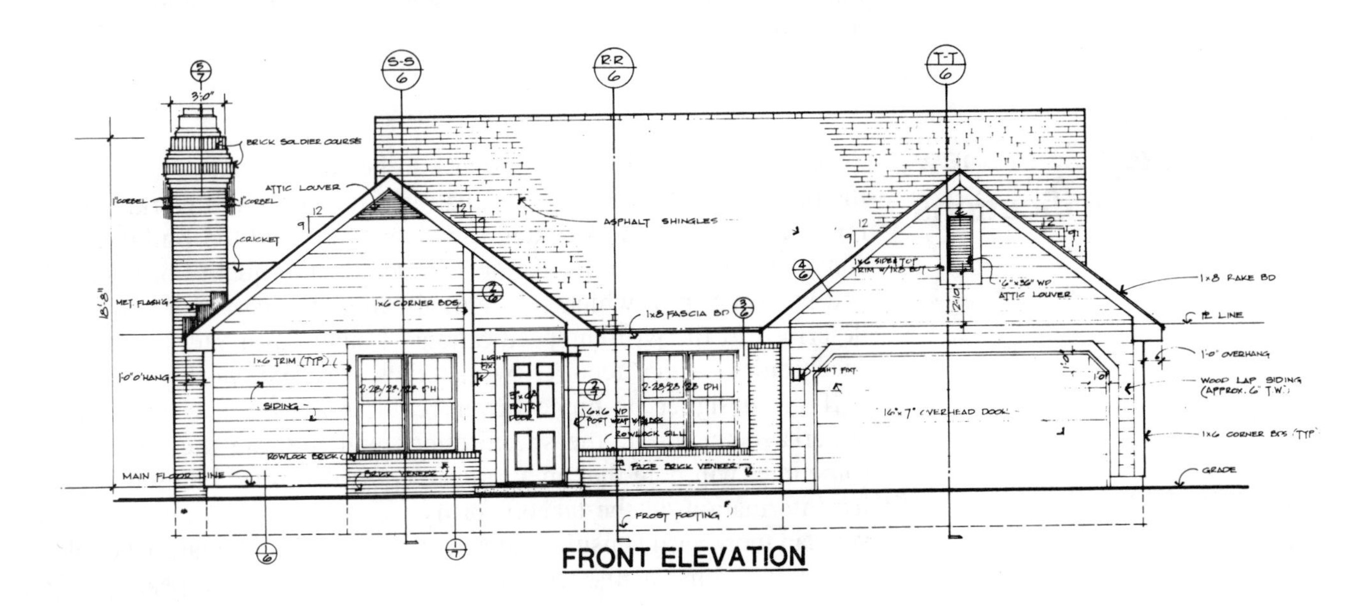

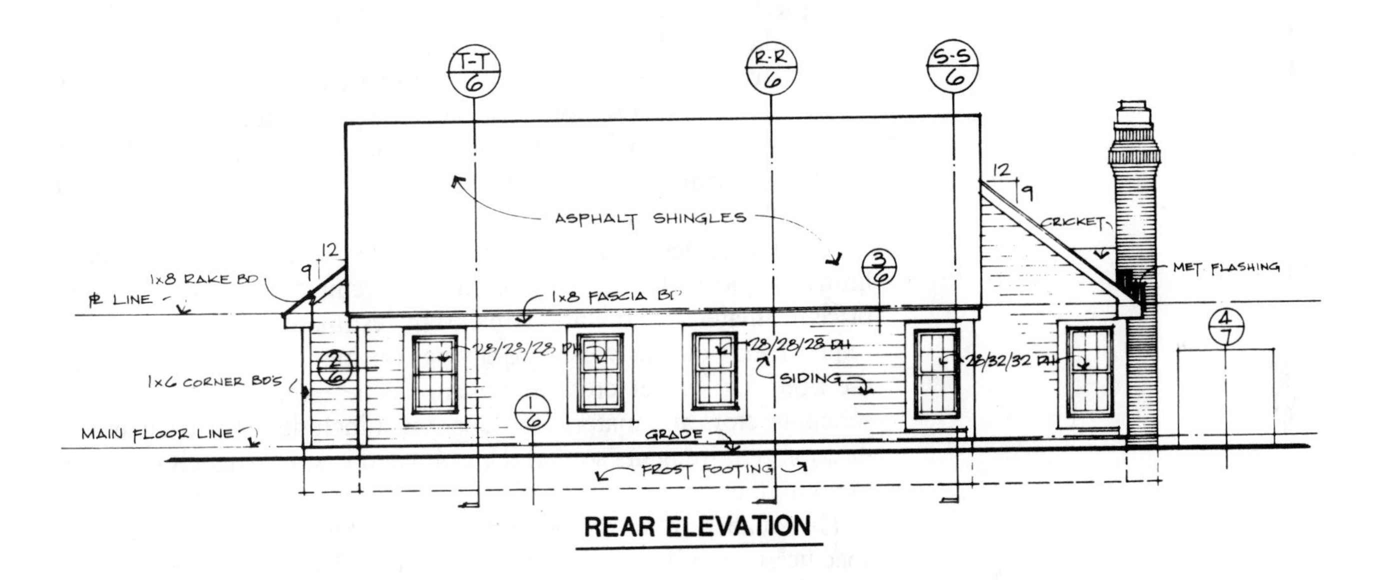

Typical Dimensions on Elevations

- Floor to cornice heights
- Window and door heights and widths
  - Roof overhangs
  - Chimney heights
  - Footing depths
- Window and door symbols and identification
- Exterior material symbols and notes
- Roof slope indication
- Special features
- Columns and posts
- Steps
- Railings
- Chimney flashing
- Exterior lights
- Gutters and downspouts
- Decorations and trim

Chapter 5, "Accuracy in Estimating," identify the standard pricing units and calculations required. Tables found in Appendix A help quantify finish items.)

Step 2. Review Specifications

Specifications describe the residence to be constructed. They explain the physical size, as well as the quality of materials and workmanship for construction. The working drawings would be too bulky if they contained all of this information. Therefore, the quality of materials and workmanship, which substantially affect the cost of a house, are described in the written specifications. Usually estimators divide the specifications into categories by materials or type of work.

If the designer or home buyer requires a particular brand of a material being used, the specifications will indicate the choice and sometimes will note that an approved equivalent may be used. If you list information on the takeoff forms according to the specifications, you can more easily ensure that prices used match the material quality and work requirements. (A sample specifications form appears in Figure 3-12.)

Typical documents may include specifications on a printed form such as those used by the Federal Housing Authority (FHA), Veterans Administration (VA), or Farmer's Home Administration (FmHA). Figure 3-12 shows a sample form filled out. Architecturally designed, custom houses often have specifications set up according to the Construction Specifications Institute (CSI) format. This format has 16 standard divisions of work.

An initial review of the specifications includes highlighting or underlining the particular information in the specifications that is needed to price the material and work items. Many estimators write specification requirements on the working drawings during the review to eliminate the need for further cross-referencing. The request for quotes from subcontractors and material vendors must include the particular specification requirements so that the estimated cost will match the cost of the purchase.

The quality of work bid must match the quality to be built. If specifications are not provided, you should fill out a specifications form, listing the materials and the quality that you intend to provide. Listing the items and the quality to be provided when the plans or specifications are incomplete helps you to avoid problems.

When you review the specifications, you should check the information listed in Figure 3-13.

Figure 3-12. Sample Specifications/Takeoff

☐ Proposed Construction

DESCRIPTION OF MATERIALS

No. ____________ (To be inserted by FHA, VA or FmHA)

☐ Under Construction

Property address Lot 2, Block C, Wood Chase S/D *City* Flat Rock *State* PA

Mortgagor or Sponsor ____________ (Name) ____________ (Address)

Contractor or Builder TLC Builders (Name) P O Box 1 Flat Rock, PA (Address)

INSTRUCTIONS

1. For additional information on how this form is to be submitted, number of copies, etc., see the instructions applicable to the FHA Application for Mortgage Insurance, VA Request for Determination of Reasonable Value or FmHA Dwelling Specifications, as the case may be.

2. Describe all materials and equipment to be used, whether or not shown on the drawings, by marking an X in each appropriate check-box and entering the information called for in each space. If space is inadequate, enter "See misc." and describe under item 27 or on an attached sheet. THE USE OF PAINT CONTAINING MORE THAN THE PERCENT OF LEAD BY WEIGHT PERMITTED BY LAW IS PROHIBITED.

3. Work not specifically described or shown will not be considered unless required, then the minimum acceptable will be assumed. Work exceeding minimum requirements cannot be considered unless specifically described.

4. Include no alternates, "or equal" phrases, or contradictory items. (Consideration of a request for acceptance of substitute materials or equipment is not thereby precluded.)

5. Include signatures required at the end of this form.

6. The construction shall be completed in compliance with the related drawings and specifications, as amended during processing. The specifications include this Description of Materials and the applicable Minimum Property Standards.

1. EXCAVATION:

Bearing soil, type Undisturbed soil - 2500 psf bearing capacity

2. FOUNDATIONS:

Footings: concrete mix 6 1/2 bag mix; strength psi 3000 Reinforcing as shown
Foundation wall: material ____ Reinforcing ____
Interior foundation wall: material ____ Party foundation wall ____
Columns: material and sizes ____ Piers: material and reinforcing ____
Girders: material and sizes ____ Sills: material ____
Basement entrance areaway ____ Window areaways ____
Waterproofing ____ Footing drains ____
Termite protection Soil poision
Basementless space: ground cover ____; insulation ____; foundation vents ____
Special foundations ____
Additional information: Monolithic slab

3. CHIMNEYS:

Material brick Prefabricated *(make and size)* integral
Flue lining: material metal Heater flue size ____ Fireplace flue size as per mfgr
Vents *(material and size)*: gas or oil heater ____; water heater ____
Additional information: ____

4. FIREPLACES:

Type: ☒ solid fuel; ☐ gas-burning; ☐ circulator *(make and size)* ____ Ash dump and clean-out ____
Fireplace: facing brick; lining firebrick; hearth brick; mantel oak
Additional information: firebrick lining integral

5. EXTERIOR WALLS:

Wood frame: wood grade, and species stud grade spruce ☒ Corner bracing. Building paper or felt N/A
Sheathing asph impg fiber bd thickness 1/2"; width 4'-0"; ☒ solid; ☐ spaced ____" o. c.; ☐ diagonal; ____
Siding redwood; grade clear; type lap; size 6"; exposure 5""; fastening galv nails
Shingles ____; grade ____; type ____; size ____; exposure ____"; fastening ____
Stucco ____; thickness ____": Lath ____; weight ____ lb.
Masonry veneer brick @ $200/m Sills brick Lintels N/A Base flashing 6 mil poly
Masonry: ☐ solid ☐ faced ☐ stuccoed; total wall thickness ____"; facing thickness ____"; facing material ____
Backup material ____; thickness ____"; bonding ____
Door sills ____ Window sills ____ Lintels ____ Base flashing ____
Interior surfaces: dampproofing, ____ coats of ____; furring ____
Additional information: galvanized flashing under siding and over rowlock
Exterior painting: material oil baced stain; number of coats 2
Gable wall construction: ☒ same as main walls; ☐ other construction ____

Figure 3-12. Specifications/Takeoff (continued)

6. FLOOR FRAMING:

Joists: wood, grade, and species ________; other ________; bridging ________; anchors ________

Concrete slab: ☐ basement floor; ☒ first floor; ☒ ground supported; ☐ self-supporting; mix ________; thickness 4"; reinforcing 6x6/10x10 wwm; insulation ________; membrane 6 mil poly

Fill under slab: material granular; thickness 4". Additional information: ________

7. SUBFLOORING: *(Describe underflooring for special floors under item 21.)*

Material: grade and species ________; size ________; type ________

Laid: ☐ first floor; ☐ second floor; ☐ attic ________ sq. ft; ☐ diagonal; ☐ right angles. Additional information: ________

8. FINISH FLOORING: *(Wood only. Describe other finish flooring under item 21.)*

LOCATION	ROOMS	GRADE	SPECIES	THICKNESS	WIDTH	BLDG. PAPER	FINISH
First floor							
Second floor							
Attic floor ____ sq. ft.							

Additional information: ________

9. PARTITION FRAMING:

Studs: wood, grade, and species stud grade spruce size and spacing 2x4 @ 16" oc Other ________

Additional information: ________

10. CEILING FRAMING:

Joists: wood, grade, and species trusses Other ________ Bridging ________

Additional information: ________

11. ROOF FRAMING:

Rafters: wood, grade, and species trusses Roof trusses (see detail): grade and species #2 syp 15kd

Additional information: trusses to be designed by registered engineer

12. ROOFING:

Sheathing: wood, grade, and species 1/2" cdx plywood 4 ply; ☒ solid; ☐ spaced ____" o.c.

Roofing ashphalt shingles; grade first; size 12x36; type 240#/square

Underlay felt; weight or thickness 15; size 36"; fastening caps

Built-up roofing ________; number of plies ________; surface material ________

Flashing: material galvanized steel; gage or weight 24ga; ☐ gravel stops; ☐ snow guards

Additional information: galvanized steel drip edge required

13. GUTTERS AND DOWNSPOUTS:

Gutters: material N/A; gage or weight ________; size ________; shape ________

Downspouts: material ________; gage or weight ________; size ________; shape ________; number ________

Downspouts connected to: ☐ Storm sewer; ☐ sanitary sewer; ☐ dry-well. ☐ Splash blocks: material and size ________

Additional information: ________

14. LATH AND PLASTER

Lath ☐ walls, ☐ ceilings: material ________; weight or thickness ________ Plaster: coats ____; finish ________;

Dry-wall ☒ walls, ☒ ceilings: material gypsum; thickness 5/8"; finish smooth

Joint treatment tape, bed, and skim (note: fire code gyp bd in garage)

15. DECORATING: *(Paint, wallpaper, etc.)*

ROOMS	WALL FINISH MATERIAL AND APPLICATION	CEILING FINISH MATERIAL AND APPLICATION
Kitchen & brkfst	wallpaper	gyp bd textured
Bath	wallpaper	gyp bd textured
Other entry	wallpaper	gyp bd textured
family rm	1/4" wd paneling - $15/sheet allow	gyp bd textured

Additional information: wallpaper allowance-$20/roll materials and labor

16. INTERIOR DOORS AND TRIM:

Doors: type six panel colonial; material white pine; thickness 1-3/8"

Door trim: type colonial; material fir Base: type colonial; material fir; size 3"

Finish: doors stain and varnish; trim stain and varnish

Other trim *(item, type and location)* ________

Additional information: ________

Figure 3-12. Specifications/Takeoff (continued)

17. WINDOWS:

Windows: type double hung; make Caradco or equal; material wood; sash thickness std

Glass: grade dbl strength B; ☐ sash weights; ☐ balances, type integral; head flashing 6 mil pol"

Trim: type colonial; material fir Paint stain and varnish number coats 3

Weatherstripping: type integral; material ______ Storm sash, number ______

Screens: ☒ full; ☐ half; type ______; number ______; screen cloth material ______

Basement windows: type ______; material ______; screens, number ______; Storm sash, number ______

Special windows ______

Additional information: insulated glass

18. ENTRANCES AND EXTERIOR DETAIL:

Main entrance door: material wood; width 3'-0"; thickness 1-3/4 Frame: material wood; thickness 3/4"

Other entrance doors: material sliding glass; width 6'-0"; thickness ___". Frame: material alum; thickness ___"

Head flashing galv matal Weatherstripping: type integral; saddles ______

Screen doors: thickness ___"; number w/slider; screen cloth material cloth Storm doors: thickness ___"; number ______

Combination storm and screen doors: thickness ___"; number ______; screen cloth material ______

Shutters: ☐ hinged; ☐ fixed. Railings N/A. Attic louvers as shown

Exterior millwork: grade and species clear redwood Paint oil stain; number coats ______

Additional information: 1/2" plywood soffit & porch ceiling, insulated glass door

19. CABINETS AND INTERIOR DETAIL:

Kitchen cabinets, wall units: material oak raised panel; lineal feet of shelves ______; shelf width 11-1/2

Base units: material oak raised panel; counter top plastic laminate; edging plastic laminate

Back and end splash plastic laminate Finish of cabinets stain and varnish; number coats 3

Medicine cabinets: make Nutone; model OU812

Other cabinets and built-in furniture ______

Additional information: ______

20. STAIRS: N/A

STAIR	TREADS		RISERS		STRINGS		HANDRAIL		BALUSTERS	
	Material	Thickness	Material	Thickness	Material	Thickness	Material	Thickness	Material	Thickness
Basement										
Main										
Attic										

Disappearing: make and model number ______

Additional information: ______

21. SPECIAL FLOORS AND WAINSCOT: (Describe carpet as listed in Certified Products Directory.)

	Location	Material, Color, Border, Sizes, Gage, Etc.	Threshold Material	Wall Base Material	Underfloor Material
Floors	Kitchen	vinyl $18/yd allowance		wood	concrete
	Bath	vinyl $18/yd allowance		wood	concrete
	Entry	vinyl $18/yd allowance		wood	concrete
	Other	carpet $15/yd allowance			

	Location	Material, Color, Border, Cap, Sizes, Gage, Etc.	Height	Height Over Tub	Height in Showers (From Floor)
Wainscot	Bath	cultured marble around tubs		60"	

Bathroom accessories: ☐ Recessed; material ______; number ______; ☒ Attached; material brass; number 8

Additional information: 2 paper holders, 2 soap dishes, 2 toothbrush holders, 2 towel bars

Figure 3-12. Specifications/Takeoff (continued)

22. PLUMBING

Fixture	Number	Location	Make	Mfr's Fixture Identificaiton No.	Size	Color
Sink	1	kitchen	Am Std or equal	M9924655	32"	stainless
Lavatory	2	baths	Am Std or equal	L5840116	18"	almond
Water closet	2	baths	Am Std or equal	W9600722	cadet	almond
Bathtub	2	baths	Am Std or equal	B9930779	60"	almond
Shower over tub △	2	baths	Am Std or equal	H4468134		
Stall shower △						
Laundry trays						

△ ☒ Curtain rod △ ☐ Door ☐ Shower pan: material ____
Water supply: ☒ public; ☐ community system; ☐ individual (private) system.★
Sewage disposal: ☒ public; ☐ community system; ☐ individual (private) system.★
★ *Show and describe individual system in complete detail in separate drawings and specifications according to requirements.*
House drain (inside): ☒ cast iron; ☐ tile; ☐ other ____ House sewer (outside): ☐ cast iron; ☐ tile; ☒ other PVC
Water piping: ☐ galvanized steel; ☒ copper tubing; ☐ other ____ Sill cocks, number 2
Domestic water heater: type electric; make and model Rheem H4382; heating capacity ____ gph. 100° rise. Storage tank: material glass; capacity 55 gallons.
Gas service: ☐ utility company; ☐ liq. pet. gas; ☐ other ____ Gas piping: ☐ cooking; ☐ house heating.
Footing drains connected to: ☐ storm sewer; ☐ sanitary sewer; ☐ dry well. Sump pump; make and model ____; capacity ____; discharges into ____

23. HEATING

☐ Hot water. ☐ Steam. ☐ Vapor. ☐ One-pipe system. ☐ Two-pipe system.
☐ Radiators. ☐ Convectors. ☐ Baseboard radiation. Make and model ____
Radiant panel: ☐ floor; ☐ wall; ☐ ceiling. Panel coil: material ____
☐ Circulator. ☐ Return pump. Make and model ____; capacity ____ gpm.
Boiler: make and model ____ Output ____ Btuh.; net rating ____ Btuh.
Additional information: ____
Warm air: ☐ Gravity. ☒ Forced. Type of system heat pump
Duct material: supply galv metal; return galv metal Insulation glass; thickness 1" ☐ Outside air intake.
Furnace: make and model York 81743 Input ____ Btuh.; output ____ Btuh.
Additional information: ____
☐ Space heater; ☐ floor furnace; ☐ wall heater. Input ____ Btuh.; output ____ Btuh.; number units ____
Make, model ____ Additional information: ____
Controls: make and types Honeywell
Additional information: ____
Fuel: ☐ Coal; ☐ oil; ☐ gas; ☐ liq. pet. gas; ☒ electric; ☐ other ____; storage capacity ____
Additional information: ____
Firing equipment furnished separately: ☐ Gas burner, conversion type. ☐ Stoker: hopper feed ☐; bin feed ☐
Oil burner: ☐ pressure atomizing; ☐ vaporizing ____
Make and model ____ Control ____
Additional information: ____
Electric heating system: type ____ Input ____ watts; @ ____ volts; output ____ Btuh.
Additional information: ____
Ventilating equipment: attic fan, make and model Nutone 4543; capacity ____ cfm.
kitchen exhaust fan, make and model Nutone 3421
Other heating, ventilating, or cooling equipment ____

24. ELECTRIC WIRING:

Service: ☒ overhead; ☐ underground. Panel: ☐ fuse box; ☒ circuit-breaker; make Round E AMP's 400 No. circuits 20
Wiring: ☐ conduit; ☐ armored cable; ☐ nonmetallic cable; ☐ knob and tube; ☐ other romex
Special outlets: ☒ range; ☒ water heater; ☐ other ____
☒ Doorbell. ☒ Chimes. Push-button locations front Additional information:
smoke detector

25. LIGHTING FIXTURES:

Total number of fixtures 14 Total allowance for fixtures, typical installation, $ $1,000
Nontypical installation ____
Additional information: ____

Figure 3-12. Specifications/Takeoff (continued)

26. INSULATION:

Location	Thickness	Material, Type, and Method of Installation	Vapor Barrier
Roof		R-30 - 9" batts plus R-5 - 1" rigid	as shown
Ceiling	2@11"	R-38 - 2 layers R-19 fiberglass batts	as shown
Wall	2@ 7"	2 layers - 3-1/2" batts	as shown
Floor			
		styrofoam vent pans as shown	

27. MISCELLANEOUS: *(Describe any main dwelling materials, equipment, or construction items not shown elsewhere; or use to provide additional information where the space provided was inadequate. Always reference by item number to correspond to numbering used on this form.)* oak fireplace mantle, 16 gage chimney cap

HARDWARE: *(make, material, and finish.)* Quickset or equal

SPECIAL EQUIPMENT: *(State material or make, model and quantity. Include only equipment and applicances which are acceptable by local law, custom and applicable FHA standards. Do not include items which, by established custom, are supplied by occupant and removed when he vacates premises or chattles prohibited by law from becoming realty.)*

dishwasher - GE 6082 or equal
range hood - Nutone 4016 or equal
range - GE 4114 Or equal

PORCHES:

as shown

TERRACES:

as shown

GARAGES:

as shown with 16'0"X7'0" Overhead door

WALKS AND DRIVEWAYS:

Driveway: width 16' ; base material earth ; thickness 4 "; surfacing material concrete ; thickness 4"
Front walk: width 4' ; material concrete ; thickness 4 ". Service walk: width ____ ; material ____ ; thickness ____
Steps: material ____ ; treads ____ "; risers ____ ". Cheek walls ____

OTHER ONSITE IMPROVEMENTS:

(Specify all exterior onsite improvements not described elsewhere, including items such as unusual grading, drainage structures, retaining walls, fence, railings, and accessory structures.)

redwood fence as shown

LANDSCAPING, PLANTING, AND FINISH GRADING:

Topsoil 4 " thick: ☒ front yard; ☒ side yards; ☒ rear yard to lot line ~~feet behind main building.~~
Lawns *(seeded, sodded, or sprigged)*: ☒ front yard ____ ; ☒ side yards ____ ; ☒ rear yard ____
Planting: ☐ as specified and shown on drawings; ☐ as follows:

2 Shade trees, deciduous, 2 " caliper.
____ Low flowering trees, deciduous, ____ ' to ____ '
____ High-growing shrubs, deciduous, ____ ' to ____ '
____ Medium-growing shrubs, deciduous, ____ ' to ____ '
____ Low-growing shrubs, deciduous, ____ ' to ____ '

16" Evergreen trees. 2 ' to 4 ', B & B.
____ Evergreen shrubs. ____ ' to ____ ', B & B.
____ Vines, 2-year ____

Figure 3-13. Specifications Needed for Pricing Material and Work Items

Concrete—Strength, additives, color, and finish for foundations, slabs, walks, steps, and drives
Masonry and Tile—Work materials
Chimneys, Flues, and Fireplace—Material requirements
Lumber—Size, grade, species, and spacing of columns, headers, lintels, joists, rafters, studs, plates, bridging, beams, girders
Flooring—Material grade and specs for subflooring and finished floors
Ceilings—Materials for subceiling, finished ceiling
Wall Materials—Grade, species, size for sheathing, siding
Insulation/Moisture Protection—Thermal insulation *R* or *U* value for walls, ceilings, roof, floors, insulation paper, vapor seal, moisture barrier
Exterior finish—Material descriptions for eave vents, soffits, fascia
Windows and Screens—For doors and frames, jamb casings
Stairs and stairways—Mechanical description and finish
Millwork—Finish, grade, and material specifications for cabinets, shelving, clothes rods
Hardware—Type and grade of finish and rough hardware items
Plumbing—Quality and type of piping and fixtures
Heating, Ventilation, and Air-Conditioning—Equipment and capacity
Painting and Decorating—Materials for exterior paint, interior paint, wallpaper

Step 3. Determine Measurements

The third step in preparing any estimate is to accurately determine all basic and critical measurements and list these individual measurements on the drawings or on the first takeoff worksheet.

One of the first lessons that a good estimator learns is to work according to a method that can be interrupted without affecting the accuracy of the estimate. Therefore, most estimators follow a fixed approach for all estimates. One method requires completing the measurements before beginning calculations and listing quantities.

An alternate method is to measure, calculate, and list the work item by item, using the plans and the checklist. Several measurements are used more than once and should be used without remeasuring or recalculating each time. Reusing measurements or calculations, however, increases the need for accurate measurement because any mistake will be compounded.

The conversion tables in Appendix A provide conversion factors to help compute material and work item quantities based on your primary measurements.

You can revise or expand the sample measurement list, (Figure 3-14) based on the type and scope of work to be performed by your crew(s) and the other materials or labor bought on a unit-cost basis.

The measurements made in this step are used in many quantity calculations required for the takeoff, so use care in measuring and recheck each item after listing it on the drawing or quantity takeoff sheet.

Figure 3-14. Basic Measurements for a Typical Residence

From the Floor Plan
Linear feet and height of exterior walls by type
Linear feet and height of interior walls by type
Number of doors by size and type
Number of windows by size and type
Linear feet of cabinets by type
Accessories (count)
Perimeter of the floor(s) with measurements
Area of floor(s) with measurements noted
Exterior dimensions within walls
Room perimeter measurements and areas

From the Foundation
Foundation dimensions (length) by type
Linear feet of footing by type
Linear feet of wall by type

From the Roof-Framing Plan or Floor Plan and Elevations
Square footage of roof
Flat area
Pitch of roof
Pitched roof area

From the Site Plan
Area of drives
Area of walks
Area of building
Area of lot

From the Room-Finish Schedule
Floor areas
Each room
Type of flooring
Linear foot of wall
Each room
List of all materials

From the Wall Section(s)
Height of walls and materials by wall type
List of materials from wall sections

Total

Item	Pricing	Unit Costs	Required Material	Labor	Equipment
________	________	________	________	________	________

Step 4. Perform the Check-Off

Using a checklist for the quantity takeoff is an efficient process of preparing a complete and accurate takeoff. Properly used, a checklist reduces the opportunity for error.

Most estimators prefer to perform takeoffs according to construction sequence. The last checklist (Figure 3-15) in this chapter is arranged this way. It groups work by material and craft classification in preparation for the cost estimate and for use in cost control. Some estimators format their takeoff checklist according to the plans rather than by construction sequence.

If the takeoff follows the sequence of the checklist, estimates will follow the same order. This arrangement allows you to stop the takeoff or cost-estimating process and resume it without omissions or counting items twice.

An accurate quantity takeoff requires that all items be included and that no item be included more than once. Each item on the checklist must be considered in each takeoff. Reviewing the working drawings before starting the takeoff allows you to eliminate consideration of checklist items that are unrelated to the structure and finishes of the house to be estimated.

To prevent double-counting, you should check off each item on the drawings as it is listed on the takeoff. Many estimators use colored pencils and a systems of *x*s and check marks to indicate that items have been listed on the takeoff. They circle dimensions on the plans, rather than mark through them to check them off, so they can still read the dimensions for other items on the plans.

Step 5. Computations

This step consists of counting, measuring, computing, and transferring the quantities of the material and the work items from the plans to the checklist. You can count many items directly from the plans and, for convenience, group like items together.

Doors, windows, and fixtures are the major cost items that are counted and then checked off on the plans. Proceed with caution when counting from foundation and roof-framing plans because these plans are not always consistent with the floor plan.

The dimensions of specific materials, such as linear feet of treated plate, determine the quantity of some items. But you can convert the measurement to pieces counted, according to a factor, such as converting the spacing of framing materials to counted pieces. However, you must group the items in the takeoff listing by material type and by different sizes or lengths required.

Example—You can convert the measurement of floor joists spaced 16 inches apart across 30 feet to counted joists by multiplying 30 × .75 and adding one (starter joist) to equal 24 joists.

Computations often involve combining dimensions and quantity factors to determine the units used for purchase.

Example—Multiply the length by the width by the depth of a slab to determine the cubic footage of concrete, and then convert that figure to cubic yards for material purchased. (The cost of finishing labor for the slab will be computed on a square-foot basis.)

You need to convert all measurements and dimensions to correct pricing units to complete the quantity determination and listing. The conversion tables in Appendix A aid in converting measurements to quantities in appropriate pricing units.

The method for determining the correct units varies in different areas and for different builders. Many builders list the pricing units on their checklists. The checklist in (Figure 3-15) includes standard pricing units arranged by work items. It more closely resembles an actual estimator's checklist than do the checklists in (Figures 3-2 and 3-3). However, your checklist arrangement should follow your personal preference.

Figure 3-15. Sample Checklist Including Standard Pricing Units

SITE WORK

Remove and replace topsoil—bank cubic yard (BCY), loose
Cubic yard (LCY)
Drain site and pumping—area
Remove or relocate culverts—lin. ft.
Remove or relocate trees—each
Board road or gravel road—sq. yd., cu. yd.
Grading, outside building—sq. ft.
Dirt fill, outside building—BCY, LCY
Buildings to move—each
Concrete walks and drives to remove—sq. ft.
Blacktop walks and drives to remove—sq. yd.
Signs and other items to remove and relocate—each
Seeding and landscaping—sq. ft.

EXCAVATION, BACKFILL, AND FILL

Foundation Excavation
Footing excavations, continuous, hand—BCY
Footing excavations, continuous, backhoe—BCY
Footing excavations, isolated, hand, machine—BCY
Pier excavations—BCY
Turn down excavations, slabs—BCY
Turn down excavations, walks—BCY
Excavations at curbs—BCY

Excess and Haul
Haul away dirt—LCY
Backfill and spread excavation excess—LCY

Fill
Dirt fill—LCY
Sand fill—LCY
Gravel fill—LCY
Flower bed fill—LCY

Grading
Rough grading, tractors—sq. ft.
Fine grading, hand—sq. ft.
Fine grading, walks—sq. ft.
Fine grading, paving—sq. ft.
Fine grading, air-conditioning pads—sq. ft.
Fine grading, at completion of job—sq. ft.

Site Excavation
Subsurface excavation, septic tanks, etc.—BCY
Trenching for utilities—BCY or LFT
Underslab compaction—LCY

Site Paving
Drives—sq. ft., sq. yd., cu. yd.
Walks—cu. yd., sq. ft.

Site Grading and Planting
Grading—sq. ft., cu.yd.
Planting—sq. ft., each

CONCRETE WORK

Division 3 of the CSI standard format specifications is the concrete check strength required. (See specifications for color, admixtures.)

Concrete Foundations
Continuous footings—cu. yd., lin. ft.
Isolated footings—cu. yd., each
Grade beams—cu. yd., lin. ft.

Concrete Slabs
Basement slab—cu. yd., sq. ft.
Slab on grade—cu. yd., lin. ft.

Concrete Walls
Basement walls—cu. yd., sq. ft, lin. ft.
Retaining walls—cu. yd., sq. ft., lin. ft.
Above grade walls—cu. yd., sq. ft., lin. ft.

Concrete Fill
Block fill—cu. ft., lin. ft.
Steps—cu. ft., sq. ft.

Sitework Concrete
Equipment pads and foundations—cu. yd., sq. ft.
Walks—cu. yd., sq. ft.
Paving—cu. yd., sq. ft.
Curbs—sq. yd., lin. ft.
Catch basins—each, cu. yd.

Miscellaneous Concrete Items
Splash blocks—each

Accessory Items
Admixtures—type, lbs.
Coloring—lbs.
Vapor barrier—sq. ft.
Runways for equipment (buggies)—lin. ft.

Concrete Forms
The cost of formwork, in square feet of contact area (SFCA), is a principal cost item involving labor and material in the concreting process. Quantities should be determined at the same time that the concrete takeoff is prepared.

Foundation Forms
Footings, isolated, continuous—SFCA, lin. ft.
Grade beams—SFCA
Walls—SFCA

Slab Forms
Slab edge—lin. ft., note edge height
Blockout (such as at depressions and offsets)—lin. ft.
Brick shelf—lin. ft. of perimeter
Screens—lin. ft.

Structural Forms
Column forms—SFCA
Sonotube form—each, lin. ft.
Beam bottom form—SFCA, lin. ft.
Beam side form—SFCA

Stair Forms
- Steps—lin. ft.
- Stair risers—lin. ft.

Site Forms
- Walks, paving, pads—lin. ft.

Miscellaneous Form Work Items
- Chamfer strips—lin. ft.
- Water drips—lin. ft.
- Key ways—lin. ft.
- Bulkheads, construction joints—lin. ft.
- Premolded expansion joints—lin. ft.
- Wood expansion joints—lin. ft.
- Metal control joints—lin. ft.
- Form release oil—gallons
- Cleaning formwork—labor only

Cement Finishing
- Grade beam, pile cap tops—sq. ft.

Slab Finish
- Slabs on grade—cu. yd., sq. ft.
- Coloring for slabs—lbs.
- Exposed aggregate slabs—sq. ft.
- Nonslip aggregate slabs—sq. ft.
- Hardened surface slabs—sq. ft.
- Broom finish slabs—sq. ft.

Site Items
- Mechanical pads—sq. ft.
- Walks—sq. ft.
- Drives and paving—sq. ft.
- Curbs—lin. ft., sq. ft.
- Exterior stairs and steps—sq. ft.

Patching and Rubbing
- Patch concrete floors, blockouts—each
- Columns—sq. ft.
- Beams—sq. ft.
- Walls—sq. ft.

Reinforcing Steel and Welded Wire Mesh
- Determine quantities at the same time that the concrete takeoff is prepared to ensure completeness.

Rebars—tons (pounds) separate by size of bar
Footings, pile caps, grade beams—lin. ft. × wt. = tons
Columns, beams above grade—lin. ft. × wt. = tons

Mesh—rolls (750 sq. ft.) separate by type

Walks, paving—sq. ft.

Slab on grade—sq. ft.

MASONRY

Concrete block (concrete masonry unit)—(8 × 16″ = 1.125/sq. ft.) Check substructure details for block.
8 × 8 × 16″—each
6 × 8 × 16″—each
4 × 8 × 16″—each
12 × 8 × 16″—each
U Blocks 12″—each
U Blocks 6″—each
U Blocks 8″—each

Clay brick
- Face brick—1,000s
- Common brick—1,000s
- Fire brick—1,000s
- Paving brick—1,000s

Clay tile
- Glazed—100s
- Smooth—100s

Mortar
- Mortar mix—bag
- Mason's sand—cu. yd.
- Mortar coloring—lbs.
- Mortar waterproofing—sq. ft.
- Parging—sq. ft.
- Drayage or handling of mortar—cu. yd.
- Portland cement—bag

Accessories
- Lintels, precast concrete, metal angles—each
- Wall reinforcing—lin. ft.
- Wall ties, brick ties—100s
- Anchors—each
- Dovetail anchors/slots—each
- Special anchors—each
- Expansion joints—lin. ft.
- Flashing—lin. ft., sq. ft.

Miscellaneous Items
- Clean and point masonry—sq. ft.
- Masonry wall sample—sq. ft.
- Waterproofing—check painting subcontractor
- Silicon wall treatment—check painting subcontractor
- Clean masonry walls with acid—sq. ft.
- Clean masonry floors—sq. ft.
- Seal masonry floors—sq. ft.

Stone
- Stone walls—sq. ft.
- Flagstone walks and floors—sq. ft.
- Stone sills—lin. ft.
- Coping stone—lin. ft.

Scaffolding—lin. ft. × height

CARPENTRY (keep treated members separate)

Prefab
- Trusses—each
- Wall sections—each

Floor Framing
- Sills—lin. ft, bd. ft., by size
- Beams, joists—lin. ft., bd. ft., by size
- Bridging—lin. ft., bd. ft., by size
- Sleepers on concrete—lin. ft., bd. ft., by size
- Subflooring—sq. ft. (10-20 percent waste)

Wall Framing
- Plates treated—lin. ft., bd. ft., by size
- Studs—each by height
- Headers—lin. ft., bd. ft., by size
- Blocking, firestop—lin. ft., bd. ft., by size
- Posts and columns—each, by size
- Exterior sheathing—number of sheets (32 sq. ft.), by thickness
- Strips to masonry—lin. ft., bd. ft., by size

ROOF FRAMING
- Joists and beams—lin. ft., bd. ft., by size
- Rafters—lin. ft., bd. ft., by size
- Ridges, hips, valleys—lin. ft., bd. ft., by size
- Collar beams, bridging—lin. ft., bd. ft., by size
- Bracing—lin. ft., bd. ft., by size
- Sheathing—number of sheets (32 sq. ft.), by thickness
- Lookouts—lin. ft., bd. ft., by size
- Purlins—lin. ft., bd. ft., by size
- Fascia backing—lin. ft., bd. ft., by size
- Roof grounds (treated)—lin. ft., bd. ft., by size
- Expansion joints—lin. ft.
- Skylight frame and trim, treated—lin. ft., bd. ft., by size
- Roof openings, for exhaust fans, air-conditioning—lin. ft., bd. ft., by size

Wallboard (Ceiling Material)
- Plywood, nailed—by thickness and finish, number of sheets, by size
- Plywood, glued—number of sheets, by size
- Sheetrock, nailed, glued—thickness, number of sheets, by size
- Sheetrock trim—lin. ft.
- Insulation board (sound, thermal control)—sq. ft., by size

MILLWORK
- Wood frames, exterior—each
- Wood frames, interior—each
- Wood doors in wood frames—each, by size
- Wood doors in metal frames—each, by size
- Hardware for wood doors—each
- Prehung doors, each, by size
- Wood windows (wood-trim windows)—each, by size
- Cased openings—lin. ft.
- Prefinished paneling—sheets
- Running trim—lin. ft.
  - Base and shoe
  - Crown (ceiling)
  - Wood handrails
  - Fascia frieze board
- Cabinets—lin. ft.
  - Base
  - Wall-hung
- Laminate countertops—sq. ft.
- Shelving—lin. ft., by size
- Closet rods—lin. ft.
- Wood flooring—sq. ft.

Window and Window Walls
- Double-hung windows—each, by size
- Window guards—each, by size
- Screens—each, by size
- Window repair—each
- Chair rail—lin. ft.

Fasteners
- Bolts
- Ramset nails
- Framing nails
- Sheetrock nails
- Plywood nails
- Concrete nails
- Masonry nails
- Roofing nails
- Air gun staples
- Adhesive for plywood
- Adhesive for wallboard
- Joist hangers
- Hangers
- Screws
- Lag bolts
- Split rings
- Columns bases
- Hurricane clips

Window Exteriors
- Siding—bd. ft., number of panels
  - 4′ × 8′ panels
  - 1′ × 4′ V-cut tongue and groove
  - 1′ × 6′ V-cut tongue and groove
- Trim—lin. ft.
- Wood roof deck—sq. ft., number of sheets
- Laminated roof deck—sq. ft., number of panels
- Soffits
- Furring (plates, studs)—bd. ft.

Caulking and Point-Up—lin. ft., check painting subcontractor
- Caulking—lin. ft., check painting subcontractor
  - Rubber
  - Silicon
  - Thiokol
- Precast panel joints
- Point-up—lin. ft., check painting subcontractor
  - at windows
  - at door frames
  - at wall cracks

Special Doors
- Hollow metal—each, by size
- Sliding—each, by size
- Garage—each, by size
- Fire doors (multifamily)—each, by size
- Dumb waiter—each by size

Chapter 4

Preparing a Complete Estimate

Once you have determined what materials and subcontracts the job will need, you can calculate the quantities. This chapter tells you how to figure out the quantities for the various work items and provides an example of a simple estimate. The breakdown shown in this example is just one way of estimating, and it may not fit your needs. Therefore, you should develop worksheets that fit your own estimating method.

The measurements recorded while reviewing the working drawings comprise the information to be listed on the worksheets. Many estimators prefer to write the basic measurements and computations on the plans so that they can use those numbers again when they list other items that require the same calculations.

Each worksheet of the takeoff should have a title block with the project name and number, so that the information will not be confused with takeoff information for other houses. Each page should be numbered, and the total number of pages shown, such as 1 of 15, so that you can determine immediately whether you have the complete takeoff and cost estimate. If the estimate has enough detail, a properly prepared takeoff can be used for material orders when you build a house.

The takeoff must be neat, legible, and well arranged so that others can use it for cost estimating, actual material purchasing, and subcontract preparation. The calculated quantities listed must be accurate—no transposed or misplaced numbers or decimal points. In listing the work items on the worksheet, if you follow the checklist, with particular attention to details and schedules, you will perform an accurate takeoff. The section and detail drawings are the sources for the quantities and materials listed on the worksheets. These drawings are usually sections of foundations, bearing and nonbearing walls, and other parts of the building that require specific explanation. These and other details are used with the checklist to calculate quantities and to list them on the worksheet in a specific order. After you have become familiar with the checklist you will use, go over the plans to be estimated and highlight the items that are not on your checklist.

The preferred formula for estimating the cost of a construction work item is as follows:

amount of work (in units)
× cost per unit
= estimated cost

Therefore, the quantity calculated in units is the information required to complete an accurate cost estimate. (Separate unit-cost factors are used for material cost, labor cost, and subcontract cost.)

Example—A wall-section shows the height between the finished floor and the finished ceiling, as well as all the materials required to build the wall. The estimator measures the length of the walls on the floor plan and checks off each material shown on the wall section as he or she lists it on the quantity takeoff. Items that have been checked off and listed on the takeoff worksheet will only be counted once. (Conversion tables for your use appear in Appendix A.)

Scanning the plans and checklists before closing out the quantity takeoff provides an accuracy check. If all items listed on the takeoff have been checked off on the plans, you can catch any omissions and add them to the takeoff. You can also correct any duplications. An experienced estimator becomes acquainted with approximate quantities of items that a house requires and will be able to detect gross errors in a single review.

The takeoff of materials must be based on standard construction practice and standard waste factors. The lap-and-waste factors included in this book are based on the authors' interpretation of locally available data. Using these factors requires judgment and utmost care. Different crews will use different amounts of materials to construct the same project. No universally applicable rules of thumb exist. Learning the traits of individual crews helps you to provide enough materials without over- or underestimating or over- or undersupplying the project.

Job Overhead

Job overhead costs are sometimes called indirect job costs because they are made up of all those costs that can be allocated to the job but that do not directly put material into place.

Example—Builder's risk insurance can be charged to a job, but it has nothing to do with any other specific cost category. The items to be considered in job overhead vary greatly depending upon the type of job. Usually the job overhead for a speculative house is greater than for a contract house. A contract house may have such jobsite overhead expenses as permit costs, inspection fees, tap fees, temporary utility costs, and a superintendent's time. A speculative house may have all these plus the cost of plans, appraisals, a construction loan, a sales commission, and sales closing costs. On a custom house, the contract should outline exactly who is responsible for what. On speculative houses, especially in an area new to you, check for tap fees, inspection fees, and other governmental and utility charges. You must also know all the closing costs for which you will be responsible, such as loan origination fees, title examination, title binders, deed preparations, deed recording fees, document preparation fees, closing fees, photos, amortization schedules, and notices of completion.

Site Work

Site work or as it is sometimes called, rough site grading, consists of work items done before the layout and construction of the house. Because many builders subcontract for the site work, they do not determine the quantities required. The site plan is the main reference for determining the work. It provides the location and size of the improvements. A site visit is usually necessary to verify the information on the plan.

Method for Determining the Quantity and the Pricing Unit

Tree Removal—Count the number of trees to be removed and list the size of the trees: 2 trees 1-foot in diameter. Check local tree ordinances in the planning stages to be sure you are in compliance.

Clear and Grub—Measure the area to be cleared in square feet or square yards.

Scrape Topsoil—Measure the area to be scraped in square feet or square yards.

Site Layout—Determine the time required based on the site contours and the footprint of the house.

Basement Excavation—To obtain the total number of cubic yards, measure and multiply: length × width × depth. To facilitate work, include an area around the actual foundation and a 45° slope of the excavation if shoring is not used.

Fill—The difference between the amount of earth you need and the amount you have is the figure you need here.

Culvert—Note the size and the length required.

Equipment Time—Many builders can estimate the number of hours of equipment time necessary to do the required sitework based on their experience on similar jobs. Because most site contractors work by the hour for residential construction, builders frequently use this method if they know the hourly rate.

Footings and Slabs

The foundation work items include the layout, excavation, and placement of the substructure for the house. A builder may subcontract a portion or all of this work. Your takeoff must supply the degree of detail you need to estimate the work based on how you do your work and how much of it you subcontract.

Figure 4-1 diagrams the foundation plan.

Method for Determining the Quantity and the Pricing Unit

Footing Excavation—To obtain the total in cubic yards, measure and multiply: length × width × depth of footing. Many estimators keep records of the time in workhours or backhoe hours required to excavate footings of certain sizes.

Backfill—Measured in cubic yards, backfill is the amount of excavated soil that must be replaced when the foundation or basement is complete. To figure the total excavation, subtract the volume of the basement, and multiply this figure by the soil shrinkage factor.

Figure 4-1. Sample Foundation Plan

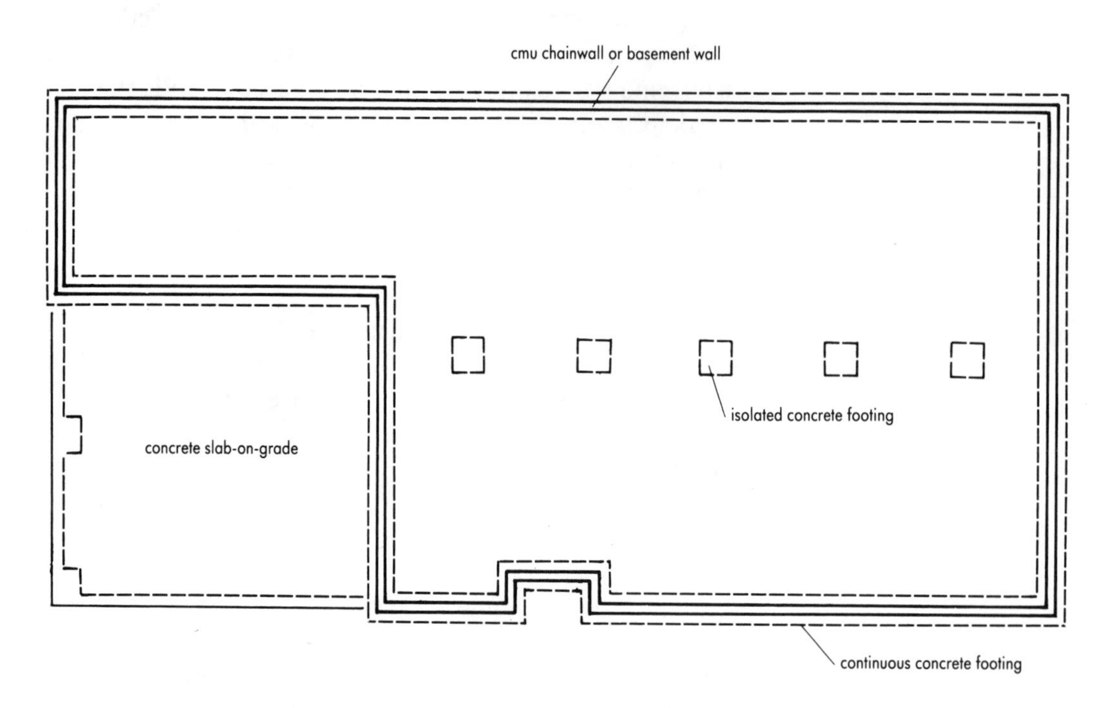

Isolated Footing—To figure the total cubic yards of concrete needed, measure and multiply: length × width × depth, and add 10 percent if you will use earthen forms.

Continuous Footing, Grade Beam—Figure the total cubic yards of concrete by measuring and multiplying: length × width × depth. Add 10 percent if you will use earthen forms.

Piers and Pedestals—Measure and multiply: length × width × depth, and multiply that figure by the number required.

Slabs for Floors—Multiply the length by the width for the total square feet of area to be finished. Multiply that number by the thickness to determine the concrete volume or use Tables 1 and 2 in Appendix A.

Walls—Multiply the square feet of wall by the thickness to obtain the cubic yards of concrete needed.

Vapor Barrier—Figure the area of on-grade concrete in square feet and add 10 percent for lap.

Batter Boards—Determine placement and lengths required from foundation plans.

Gravel—Use the square-foot area of the slab less the area taken up by footings. Usually the gravel is 4 inches thick, and a yard of gravel covers 81 square feet. To get the number of tons, multiply cubic yards by 1.6.

Rebars—Continuous bars must have lap joints. Check the length of the lap required in the specifications for bar diameters and length of bar to be ordered—usually 20, 30, or 40 feet. It is measured in linear feet and ordered in tons or pounds (Table 4, Appendix A).

Mesh—Figure the area of the slab in square feet. Each roll contains 750 square feet, but figure coverage at 700 square feet per roll to allow for lap.

Dowels—Order these by piece length plus spacing, plus one for each corner.

Anchor Bolts—Count these bolts based on spacing and number of pieces and note length and diameter.

Formwork—Measure the edge of the slab to obtain the linear feet and height. Multiply these to obtain the square feet needed.

Screeds—Draw these temporary forms on the slab or the foundation plan and measure them in linear feet. (They ensure level placement of the concrete slab.)

Masonry

Because of its simplicity and accuracy, the most widely used takeoff method for masonry is the wall-area method. The steps for this method are listed below:

1. Determine the gross area (square feet) of the wall.
2. Total the area taken up by openings.
3. Multiply the total openings area by 80 percent to allow for cuts and returns.
4. Subtract the net openings area from the gross wall area.
5. Multiply the net wall opening by the appropriate factor in Table 11 in Appendix A to determine the number of units needed.

Accurately recording the information derived provides quick reference for related accessory items. Accepted practice lists both brick and block with the thickness first, followed by the face dimensions (height and width).

The takeoff of masonry units must include factors for waste, cutting, and breakage during shipping. The combined loss factor can vary from 3 to 12 percent, with 5 percent as a reasonable estimating factor in most situations. The total number of units must be rounded up to the next purchase-unit number (usually around 500 for bricks). Figures 4-2 and 4-3 diagram masonry foundations.

Method for Determining the Quantity and the Pricing Unit

Masonry Units—Order bricks by the cube (usually 500), blocks by the total number, and stone by square feet (Tables 11 and 13, Appendix A). To determine gross area of walls, multiply length by height. The net area equals the gross area less 80 percent of the opening area:

$$\text{net area} \times \text{waste factor} = \text{number of bricks or blocks}$$

Columns—Count the blocks from the detailed drawings. For bricks add the volume of the column to the volume of brick (nominal size).

U Block—For each bond beam, multiply the perimeter by .75 for the number required. For the number required for each lintel, add 16 inches to the width of the opening and multiply by .75. Note that these blocks would have been included in the number determined for the entire wall area in a previous calculation and must be deducted from that number to provide an accurate count of the standard blocks required.

Figure 4-2. Sample Chainwall Foundation

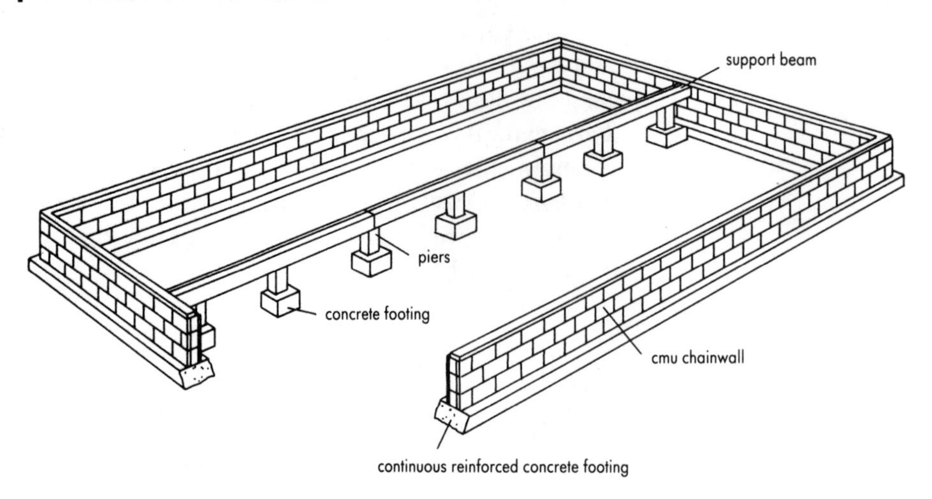

Figure 4-3. Sample Basement Walls and Foundation

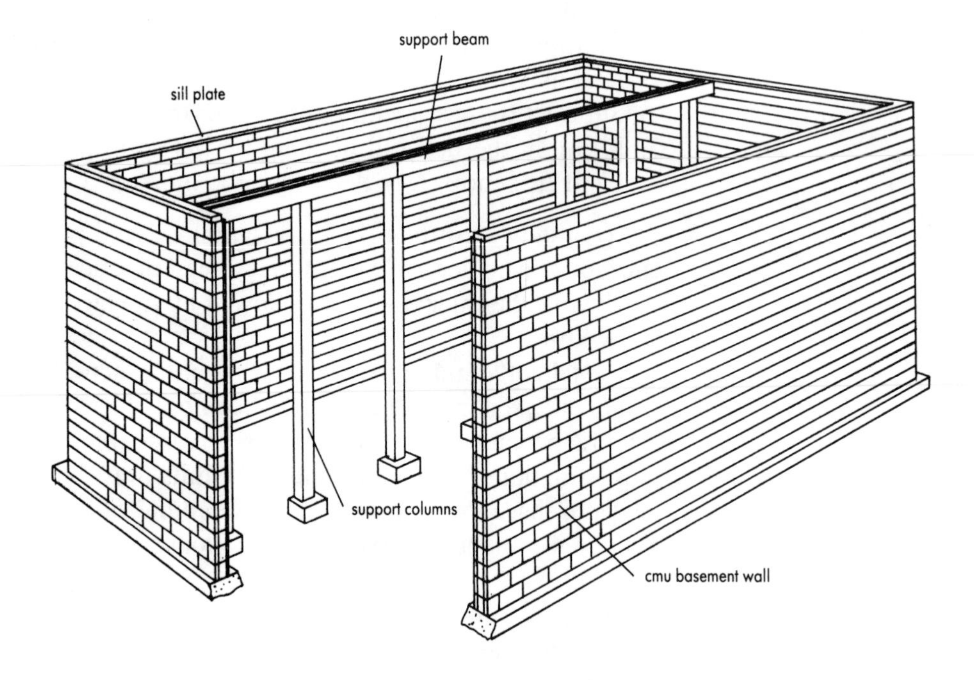

Horizontal Joint Reinforcement—Measure in linear feet to figure the amount needed (Table 12, Appendix A). Truss and ladder reinforcing is available in 10-foot lengths requiring 6-inch laps. Multiply the total linear feet by 5 percent to account for waste.

Wall Ties—If you multiply vertical spacing in inches by horizontal spacing in inches and divide by 144 you will get the number of square feet per tie. Divide this result into the total sum for the number required.

Mortar—Most estimators use a factor of 140 bricks per sack and a factor of 33 standard blocks per sack (Tables 11 and 13, Appendix A.)

Mason's Sand—Ordinarily 1,000 bricks take about ¾ yard of sand. However, to allow for waste, most estimators figure 1 yard per 1,000 bricks or 1 yard per 7 sacks of mortar.

Blockfill—(For *U* block and block cell, see Table 14, Appendix A.)

Rough Carpentry

The framing or rough carpentry usually takes more time to estimate than any other category. Figure 4-4 shows the basic elements of framing.

The wall sections provide the dimensions from the finished floors to the finished ceilings in addition to the size and spacing of framing members. The cornice portion of this detail helps to determine the framing and backup members required where the roof is attached to the walls.

Framing lumber is available in lengths of 8 feet or more, in multiples of 2 feet. You must allow for loss from cutting when taking off lumber for estimates or material orders. Many estimators prefer to round off to the next highest multiple of 2 feet when they list framing lumber on the takeoff.

The cost of lumber per board foot increases substantially for lengths greater than 16 feet. Therefore, you should list this premium-length lumber separately on the takeoff for pricing the lumber.

Example—An estimator must take off 2x4s for a 9-foot wall. If precut 9-foot studs are unavailable, the takeoff can list 10-foot pieces or 18-foot pieces for two studs. The estimator compares the material cost for 100 studs as follows:

50 18-ft. studs = 900 lin. ft. = 675 bd. ft. × $/bd. ft. = ________
100 10-ft. studs = 1,000 lin. ft. = 750 bd. ft. × $/bd. ft. = ________

If labor is bought on an hourly basis, the estimator would add the labor cost for cutting the framing members and use the more economical approach, provided the lumber would indeed be bought and used that way. A checklist for the rough carpentry takeoff appears with the drawing of the rough carpentry in Figure 4-4.

The Floor-Framing Takeoff

The takeoff method for wood-framed floors is the same for floors built on piers or crawl spaces, above basements, and for second-floor framing. The measurements required include the perimeter, length, and width of the floor, the span(s), and spacing of support members. These items are discussed in the paragraphs below and they are visible on the floor-framing plan (Figure 4-5).

Method for Determining the Quantity and the Pricing Unit

Listing all treated lumber separately on the takeoff will help to ensure accurate pricing.

Sills—Measure the perimeter (outside) of the house in linear feet and check for treated lumber.

Beams—To obtain the linear feet of beam lumber needed, round the beam length and bearing length up to the next numbers divisible by 2:

beam length
+ bearing length
= total of beam and bearing length
× number of boards per beam
= linear feet of beam lumber

Ledgers—Multiply the beam length by 2, if these are required.

Joists—To determine the number of spaces, multiply the dimensions perpendicular to the span of the joist by the spacing factor (Table 5, Appendix A). And add one additional joist to complete the floor framing. If the joists are 16 inches on center, the joist count is the perpendicular dimension multiplied by .75 plus 1. If the joists are 24 inches on center, multiply by .50 and add 1.

Example—If a span is 14′7″, the perpendicular measurement 26′, and spacing 16″ on center:

$$28' \times .75 = 20 \text{ spaces} + 1 = 21 \text{ joists } 16' \text{ long}$$

You add one more joist for each nonbearing partition that will be parallel to the joists, and add two joists for each parallel bearing partition.

number of joists
× length of the joists (2′ measurement)
= linear feet of joist material required

Bridging—Joist spans of less than 14 feet require one row of bridging; spans over 14 feet, two rows:

dimension perpendicular to the joist span
× number of rows of bridging
= linear feet of bridging

For cross bridging multiply the total in this transaction by 3.

Headers—To determine the linear feet of header material needed to box in the ends of the joists and to frame openings in the floor, multiply the dimension perpendicular to joist span by 2. For openings, measure their perimeters and round up to the next even number for linear feet. Openings wider than 4 feet require double headers, so multiply those by 2. Headers are usually the same size as joists unless the plan specifies otherwise.

Subflooring—The floor area measured to the outside of the stud line of the exterior wall is the subflooring area. If the floor has a number of openings in the floor, subtract the area of the openings if they are significant (over 32 square feet). Table 6 in Appendix A provides multipliers for various types of subfloors. If the floor material specified is plywood or particle board, a reasonable practice is to round the length and width of the floor up to the next 2-foot measurement before determining the area to allow for cuts and waste. The quantity is listed in full 4 × 8-foot sheets.

The Wall-Framing Takeoff

The measurements required for the wall-framing takeoff are the total length of exterior and interior walls, measured from the floor plan. Figure 4-6 diagrams the wall framing. The wall-section drawings also provide specific details and information for the framing. (See Figures 3-8, 4-7, and 4-8.)

Wall-framing members are usually 2x4s, double 2 × 4s, 2 × 6s, or 2 × 8s, and the plates are the same size as the studs. A single plate on

Figure 4-4. Sample Rough Carpentry

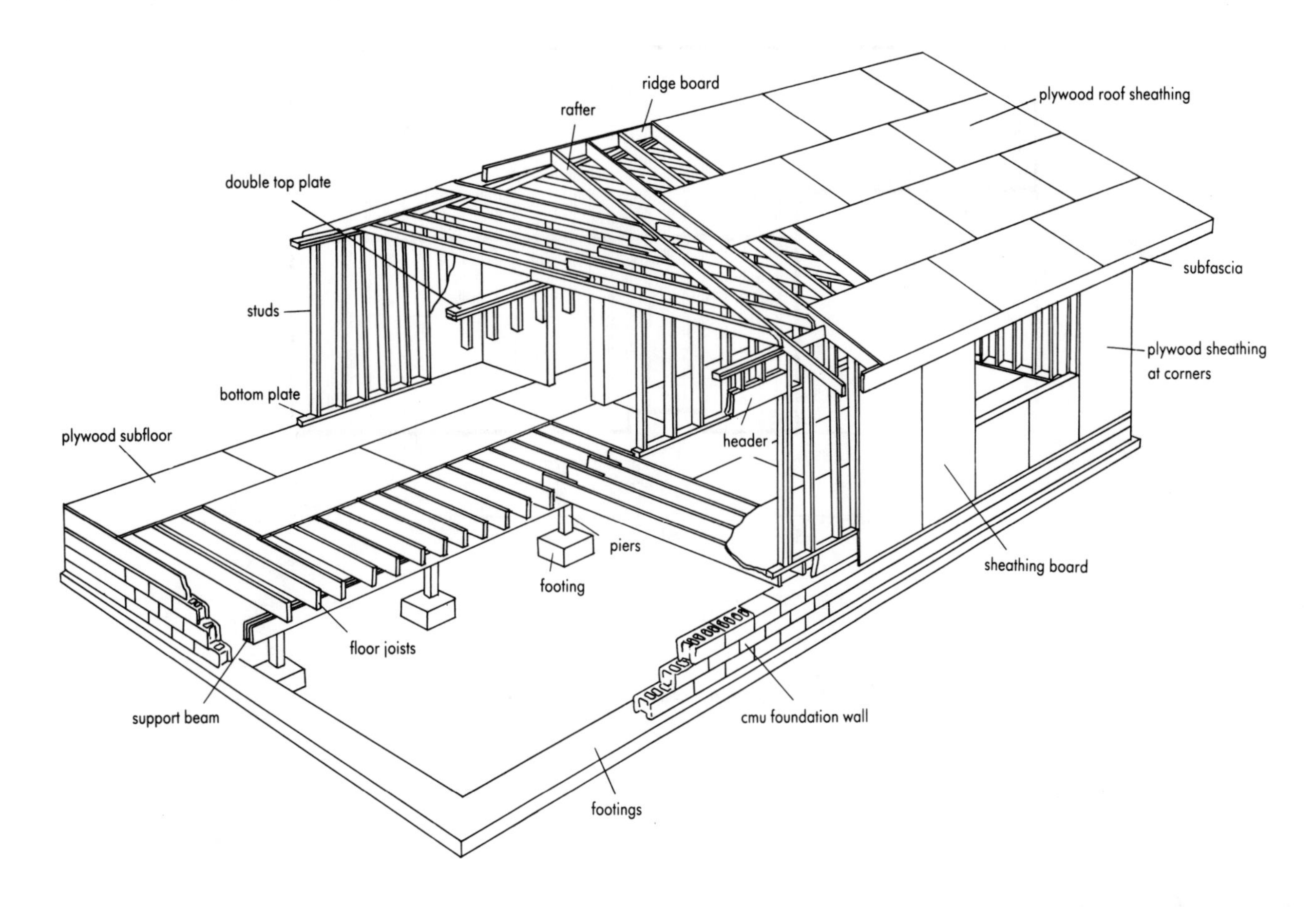

Checklist for the Rough Carpentry Takeoff

Floors
Sills
Beams
Joists
Bridging
Headers at openings, blocking
Subflooring (building paper)
Ledgers
Screeds

Walls
Plates
Studs
Headers
Gable framing
Blocking, firestops
Nailers, girths
Bracing at corners
Exterior sheathing
Posts, columns
Beams
Furring
Backing for trim

Roofs
Ceiling joists
Bridging
Beams
Headers at openings
Rafters
Ridges
Hip-and-valley members
Trusses
Leveling beam (strong back)
Collar beams
Bracing
Lookouts
Dormers
Purlins
Roof sheathing
Backing for trim

Stairways
Treads
Risers
Stringers
Carriage boards
Nosing
Disappearing stairways

Figure 4-5. Sample Floor Framing

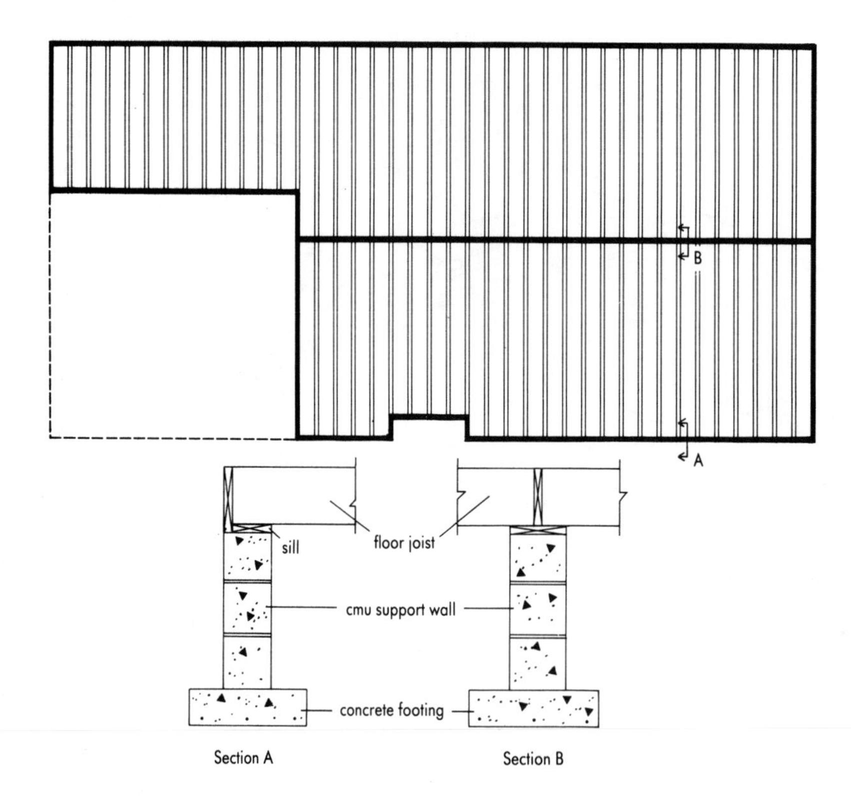

Figure 4-6. Sample Wall Framing

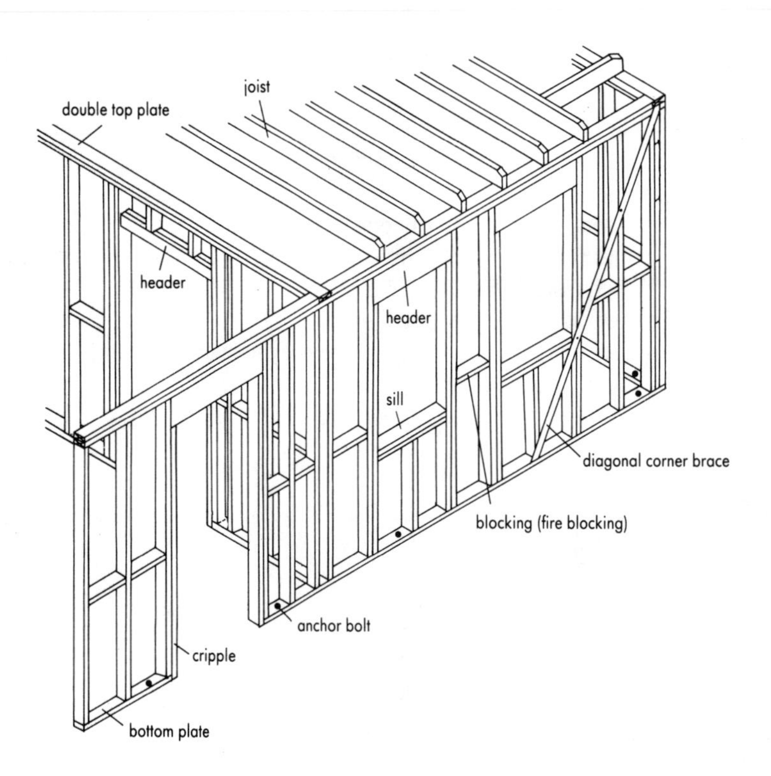

Figure 4-7. Sample Wall Section

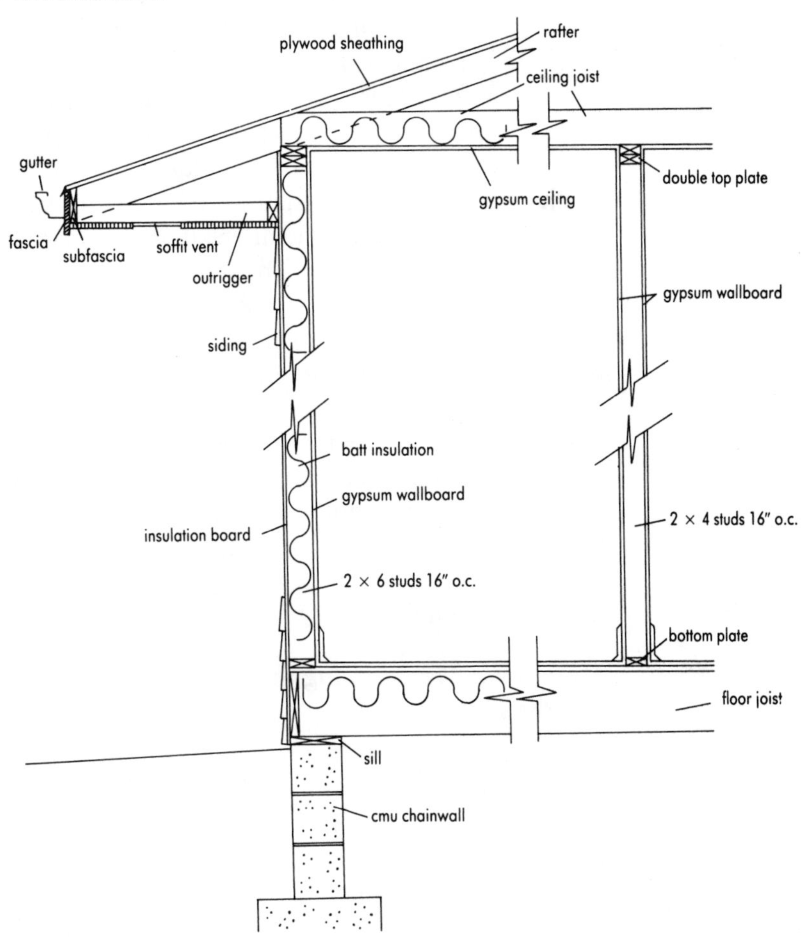

Figure 4-8. Sample Exterior Brick Veneer Wall Section

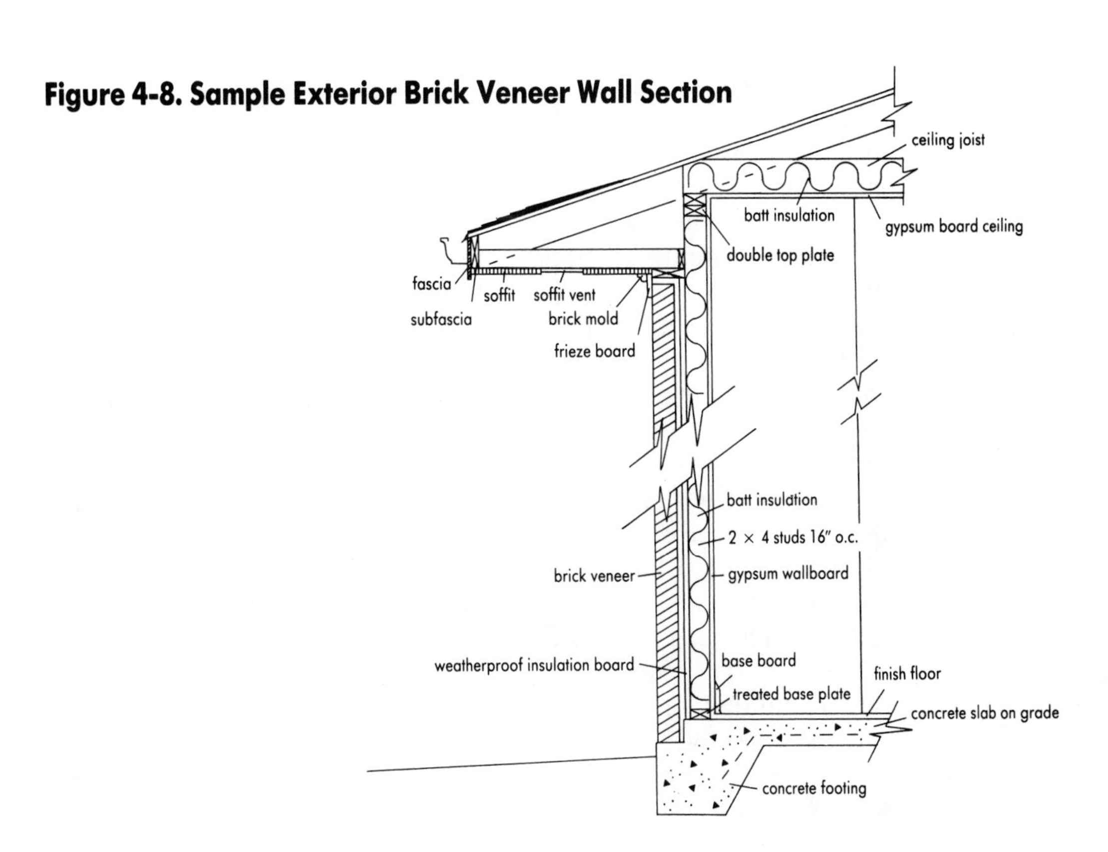

the bottom and a double plate on top is standard. The bottom plate is treated wood when it is attached to concrete or masonry.

In some instances the sizes of the headers and beams are not on the plans. Many builders prefer to determine the largest member (header or beam) that is needed and make all headers the same size to simplify the field work. In some cases, builders cut and use dimension lumber (such as 2 × 12s used for slab formwork) for headers and beams.

Many of the framing members are ordered in specific lengths, such as precut, squared studs. The quantity takeoff lists the linear footage of material required on the worksheets. Some estimators and suppliers prefer to price lumber in board feet. If conversion to board feet is required, it is the last step in taking off the lumber. Various lengths of a particular dimension of lumber are grouped in the takeoff if they are priced the same per board foot (Figure 4-9.)

The framing takeoff should segregate the lumber by size (nominal dimension), length, and location of use. Treated or other special-cost material must be listed separately. The final listing of material can be formatted as follows:

Size	Length	Quantity	Total Length	Board Feet

Method for Determining the Quantity and the Pricing Unit

Plates—For the standard single bottom plate and double top plate, multiply the total linear feet of wall by 3 to obtain the total linear feet needed. The bottom plate must be treated if it is placed on a concrete slab, and it must be kept separate from the top plates on the quantity takeoff because of the cost difference of the treated member. Door openings are usually not deducted when determining the bottom plate to account for loss from cuts.

Studs—Builders estimate studs several ways. When the studs are spaced at 16 inches on center, one stud per linear foot of wall plus one stud for every corner, tee, door, and window usually takes care of the material required. Alternately, you can multiply the length of the wall by .75 to establish a stud count and add two studs for every corner, tee, door, and window.

Headers—Use the schedule of doors and windows to determine the width of each opening (the span of the header). Multiply the width of each door and window opening by the number of openings required. Add 8 inches to provide for header bearing, 2 ends × 4 inches. The sum of all opening widths plus bearing length equals the total linear feet of material required. The total linear footage must be multiplied by 2 for double headers (Figure 4-10).

Blocking, Firestops—For one row of blocking, the blocking equals the linear footage of the wall; for two rows, double the linear footage. Miscellaneous blocking for cabinets, shelves, and rods depends on the method of fastening. Measure the linear footage of cabinets or items requiring blocking and multiply the total linear footage by 2.

Bracing—To determine the diagonal bracing needed at exterior corners multiply the wall height by 1.42 and multiply that figure by 2 at each corner for 2 plywood sheets per corner.

Furring—Furring above cabinets and for false ceilings is a short wall that has plates and studs so it is taken off as a separate wall based

Figure 4-9. Sample Wall-Framing Takeoff

WORKSHEET : WALL FRAMING

	ITEM	MEASUREMENT		QUANTITY	MATERIAL UNIT COST		MATERIAL COST
1							
2	EXTERIOR WALL 2x6	196 LFT					
3							
4	INTERIOR WALL 2x4	204 LFT					
5	2x8	16 LFT					
6							
7	TREATED PLATE						
8	R/L 2x4			204 LFT			
9	2x6			196 LFT			
10	2x8			16 LFT			
11							
12	TOP PLATE						
13	R/L 2x4			408 LFT			
14	2x6			392 LFT			
15	2x8			32 LFT			
16							
17	STUDS PRECUT						
18	8' 2x4			210 PCS			
19	2x6			200 PCS			
20	2x8			20 PCS			
21							
22	BLOCKING, FURRING						
23	SUBFASCIA, SOFFIT						
24	NAILER, FRIEZE NAILER						
25	R/L 2x4			1000 LFT			
26							
27	HEADERS						
28	DOORS (WIDTH)	52 LFT	2x8	120 LFT			
29	WINDOWS (WIDTH)	62 LFT	2x8	130 LFT			
30							
31							
32	EXTERIOR SHEATHING						
33	5/8" BOARD 2'x8'	196 LFT (CORNERS)		90 PCS			
34							
35	PLYWOOD CORNERS	4		8 SHTS.			
36							
37							
38							
39							
40							

Figure 4-10. Figuring Header Material

Number	Opening Width	Double Header Length	Material
1	6′	6′8″	13′4″
3	4′	4′8″	28′
3	2′8″	3′4″	20′
			61′4″

on the length of the shorter studs (in 2-foot measurements). These walls often have a single rather than a double top plate.

Exterior Sheathing—The linear footage of the exterior wall multiplied by the height of the wall equals the gross area. For plywood or fiberboard, subtract half of the area of the door-and-window openings from the gross area and divide the result by the square footage of the sheathing panel:

32 sq. ft. for 4x8s
16 sq. ft. for 2x8s
36 sq. ft. for 4x9s

Sheathing is not required where plywood is specified for corner bracing, and you can subtract the number of corner sheets from the number of sheets calculated.

On houses with gable roofs, the ends of the attic must be framed to enclose the house. In many cases the estimator has no plan or detail of this framing for the house being estimated. Some estimators sketch the framing on the elevations or on sketch paper to have a clear understanding of the materials needed (Figure 4-11 and 4-12). You should use the roof-framing plan (Figure 3-10) and the elevations (Figure 3-11 to count the gable walls to be framed.

Gable Studs—Divide the width of the structure in half, multiply that figure by the spacing factor (.75 for 16 inches on center) for the number of studs required. The length of each member equals the roof rise rounded off to the next highest, even number of feet. For example, if the roof rise is 6 feet on a house that is 36 feet wide and built with a 4 on 12 pitch roof, the estimate of studs for the gable wall is

36′ × .5 × .75 = 13.5 (or 14 pieces, each 6′ long

Cutting 6-foot pieces to the appropriate stud length for half of the gable wall produces the appropriate length of studs for the other half.

Nailers—To figure the amount of linear feet required:

half the width of the building
× spacing factor (.5 for 2-foot spacing)
= total of half the width × spacing factor
× roof rise
= the linear feet of horizontal nailers

Figure 4-11. Sample Gable Wall Framing

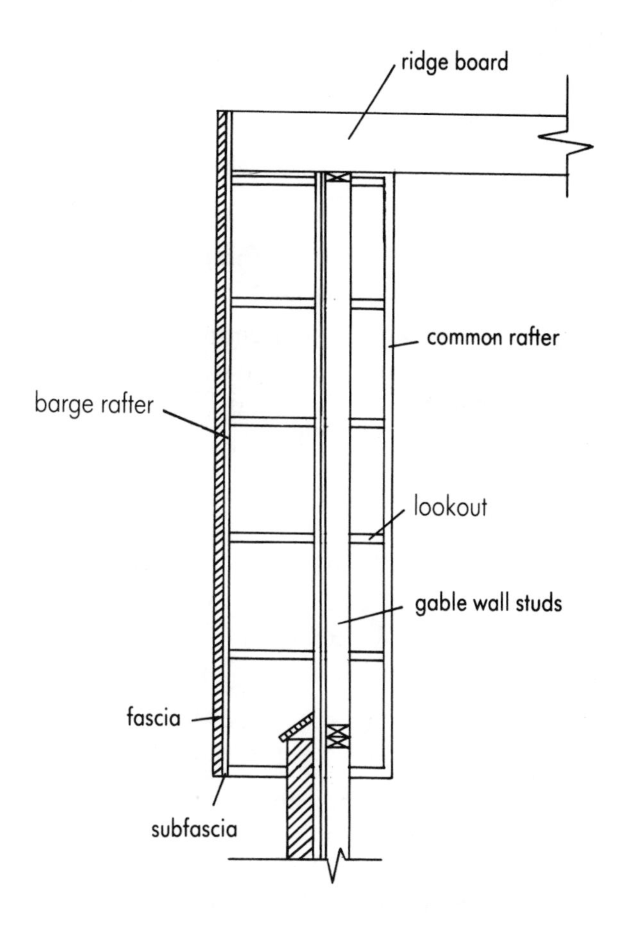

Figure 4-12. Sample Gable Roof Framing

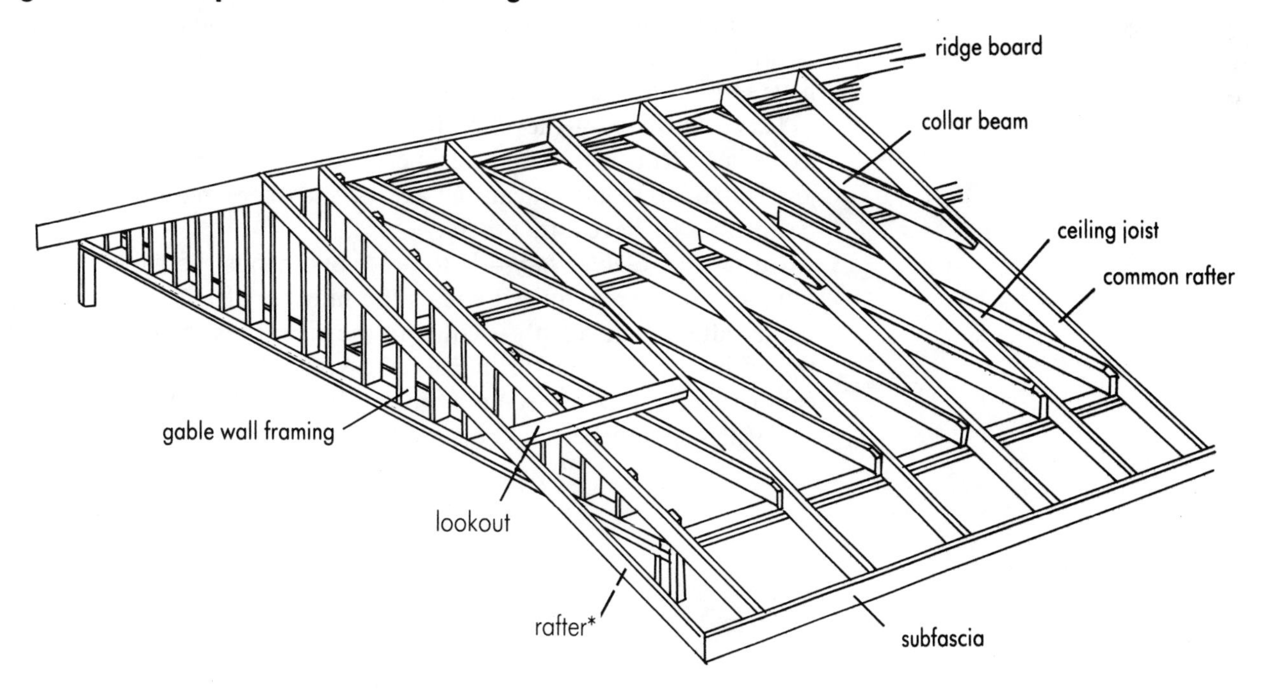

*Also called fly rafter and rake rafter.

The Roof-Framing Takeoff

Measuring the roof or doing the computations of materials needed for the roof-framing plan using the roof-rafter multipliers (Table 7, Appendix A) provides the basic information for the takeoff of roof-framing material. Figure 4-13 provides the roof-framing takeoff for a gable roof with a 2-foot overhang.

Trussed roofs are simpler to take off than are stick-framed roofs. Likewise, gable roofs are easier to take off than are hip roofs.

Method for Determining the Quantity and the Pricing Unit

Trusses—To determine the number of trusses required, multiply the dimension perpendicular to the direction of the truss span by the spacing factor and add 1 truss for the end (Figure 4-14.) The spacing factors are as follows:

.75 for 16′ oc
.50 for 24″ oc

Joist—Multiply the dimension perpendicular to the span of the joist by the spacing factor to determine the number of spaces and add 1 joist for the total required. The length of the member(s) equals the span rounded off to the next highest even footage:

16″ oc—perpendicular dimension × .75 + 1 = joists
24″ oc—perpendicular dimension × .50 + 1 = joists

Joists seldom span the entire width of a house; therefore, the actual length and number of pieces required must be carefully calculated and counted. Sketching the joist-framing layout on the floor plan helps to prevent errors (Figure 4-15).

Beams—To determine the linear footage of material required, add the beam length and the bearing length (usually 4 inches each end) and multiply by the number of boards per beam (usually 2). The spans of the beams determine the size(s) of the framing members. You may have to refer to the floor plan to locate and size the of beams.

Leveling Beam—A leveling beam may be required if the joist span exceeds 12 feet. The dimension perpendicular to the direction of the joist is equal to the linear footage of members required. Leveling beams usually have 2 members.

Rafters for Gable Roof—If the roof overhangs the gable ends, the number of rafters required equals the number of joists required plus barge rafters. The length is equal to the span plus the overhang multiplied by the rafter factor in Table 7, Appendix A. The length of the common rafters is the same as that calculated for gable roofs. Hip-and-valley members must be added to roof-framing takeoff.

Figure 4-13. Sample Roof-Framing Takeoff

ROOF FRAMING TAKE-OFF

	ITEM	MEASUREMENT		QUANTITY	MATERIAL UNIT COST		MATERIAL COST
1							
2	TRUSSES 6/12						
3	2x4 CONSTRUCTION						
4	28' SPAN w/2' O'HANG	40'	2' O.C.	21 EA			
5							
6	PLYWOOD SHEATHING			48 SHTS			
7	5/8"						
8							
9							
10	ALTERNATE: STICK FRAME						
11							
12	JOISTS 2x6	16'		21 pcs			
13		14'		21 pcs			
14							
15	RAFTERS 2x6	20'		42 pcs			
16							
17	BRACING 2x4	R/L		250 LFT			
18							
19	PLYWOOD SHEATHING						
20	5/8"			48 SHTS			

Figure 4-14. Prefabricated Truss

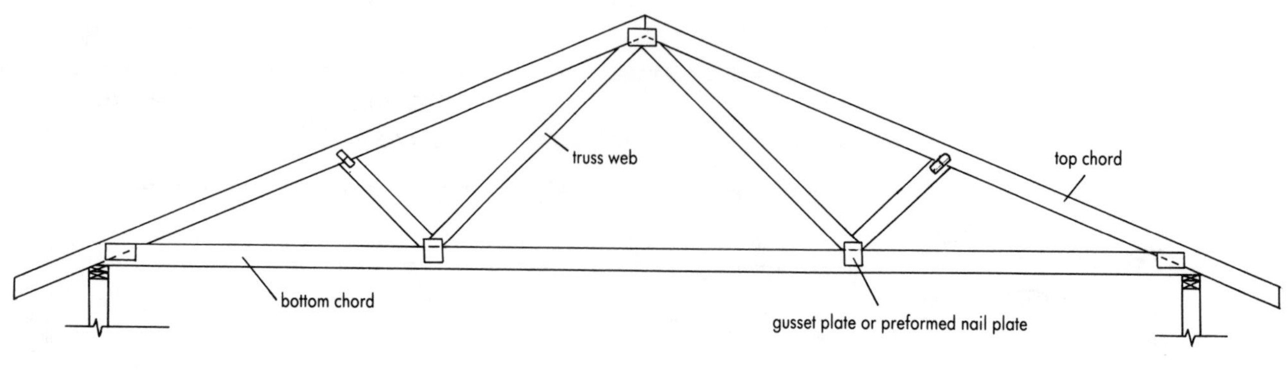

Figure 4-15. Sample Rafter-and-Joist Roof Framing

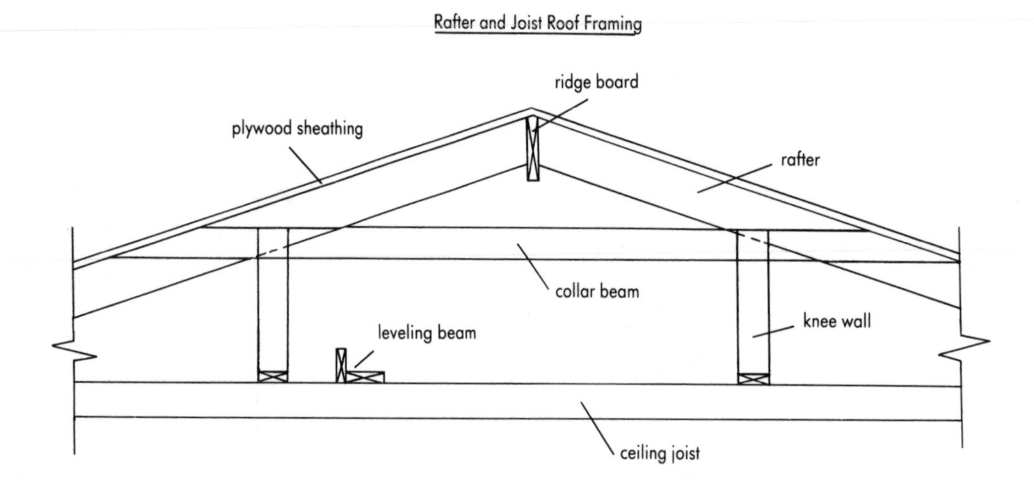

Jack Rafters—A hip roof requires jack rafters (Rafters that are shorter than usual.) to complete the framing (Figure 16). You can view these as cut common rafters. The number of common rafter lengths equals the count of common rafters for a gable roof the same size plus 2 (one on each end).

Hip-and-Valley Framing—Determine the length of the hip-and-valley members by multiplying the flat (diagonal) measurement of the member by the hip-and-valley rafter factor in Table 7, Appendix A. (See also Figure 4-17.)

Ridges for Gable Roof—Use the building dimension parallel to the ridge and add twice the overhang, or measure the ridges on the elevation in linear feet.

Figure 4-16. Sample Hip Roof Framing

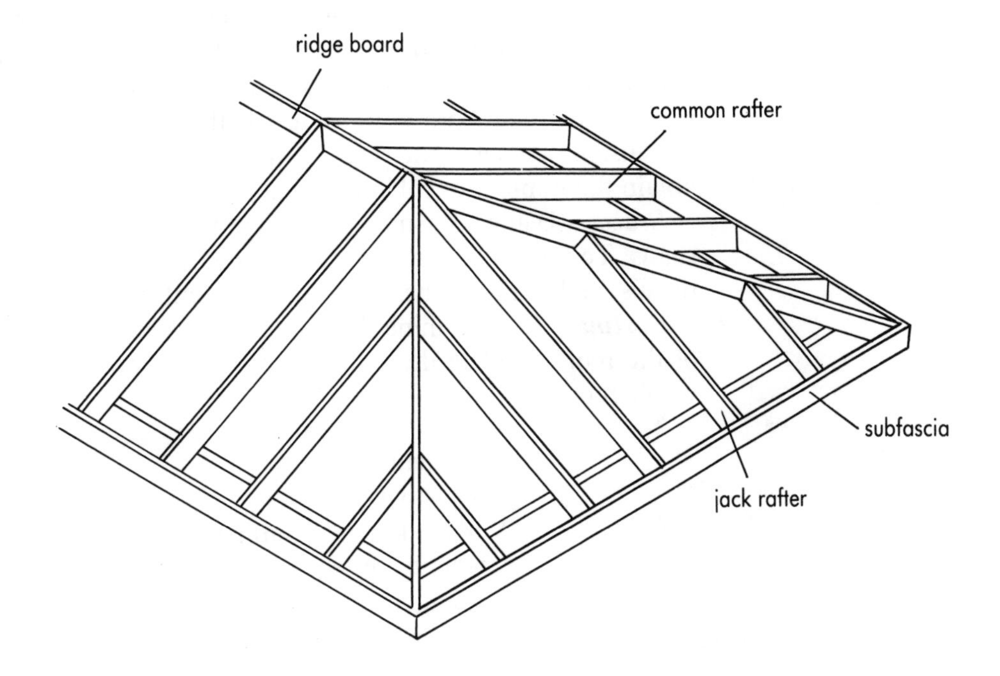

Figure 4-17. Sample Hip-and-Valley Roof Framing

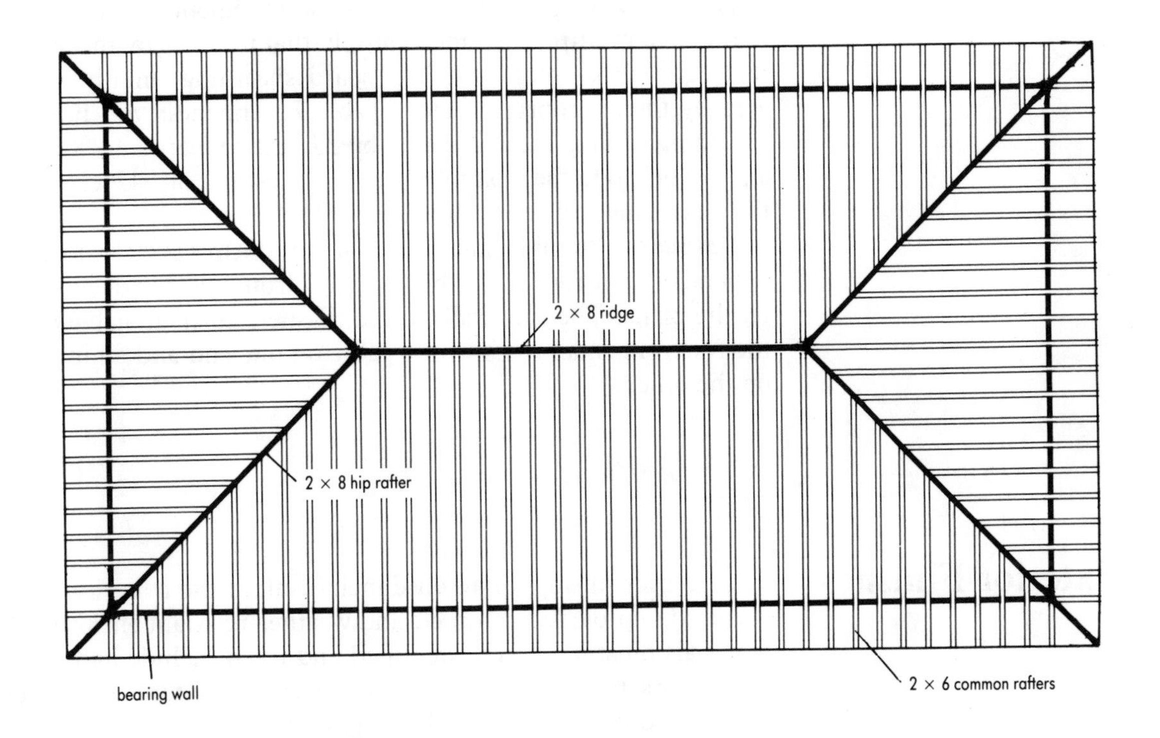

Ridges for Hip Roof—Use the *building* dimension parallel to the ridge and subtract the building dimension perpendicular to the ridge, or measure linear feet of the ridges on the elevation (Figure 4-17).

Collar Beams—When trusses are not used, collar beams are required at 4 feet on center; 10 feet is the usual length. Multiply the dimension parallel to the ridge by the spacing factor (.25 for 4 feet) to obtain the number of collars needed.

Braces—Using these members can reduce the size of rafters for large spans. To find the linear feet needed, measure bracing members from details, and multiply the length by the number of members required.

Purlins—To determine the number of pieces required for one side of the roof complete the transactions below and multiply by 2 for a whole roof:

rafter length
× purlin spacing (.5 for 24″ oc)

= *X* (rafter length × purlin spacing)
+ 1 starter piece

= pieces for 1 side of roof
× purlin dimension parallel to ridge (1 × 4 or 1 × 6)

= linear footage for 1 side of roof

Headers—The perimeter of the opening plus 30 percent of that amount (rounded up to 2 feet) for each framing member required for skylights, vents, or other roof openings.

Roof Sheathing—Round the flat measurement of the roof area up to the next multiple of 2 and multiply that number by the area factors that equal gross roof area. For board sheathing, multiply the gross area by the appropriate factor (Table 7, Appendix A). For plywood sheathing, divide the gross area by 32 square feet (a 4 × 8-foot sheet) and round up to the next number. Many estimators sketch a layout on the elevations and count the number of sheets.

Dormers—Framing and material for dormers and miscellaneous appendages on roofs must be taken off from specific details provided on the plans. The side walls of the dormer are small gable walls, and the roofs are small roofs with a gable on one end and a hip and valley on the other.

Exterior Finishes

The exterior finishes of a house include all of the materials and items required to provide a complete and weather-resistant enclosure. Those materials must be clearly identified and often include items that subcontractors might supply, such as electrical fixtures, chimney/flue, gutters and downspouts, and skylights.

Be sure that items shown in exterior wall sections and the roof and attic sections that are not included in the framing takeoff are included with the exterior or interior finishes. Batt and/or board insulation, vapor barriers, and other items providing protection from the elements must be included in the quantity determination and takeoff.

Elevations

The elevations provide a flat, correctly scaled picture of the front, rear, and sides of a house. Measure and compare the elevations to the dimensions on the floor plan to complete the quantity determination. Details, including cornice sections and wall details, are required to specifically describe the size and dimension.

Method for Determining the Quantity and the Pricing Unit

Exterior Sheathing—Multiply the linear footage of exterior wall by the height of the wall for the gross wall area. For plywood, subtract 50 percent of the opening area to determine net area, and divide by 32 square feet per panel to determine the total number. For fiberboard, use gross area, less 5 percent of the gross area, and divide by square footage of panel size:

32 sq. ft. for 4′ × 8′
16 sq. ft. for 2′ × 8′
36 sq. ft. for 4′ × 9′

Less sheathing is needed if plywood has been specified for corner bracing.

Exterior Siding or Paneling—Multiply net wall area by the appropriate factor in Table 6, Appendix A.

Paneling and Vertical Siding—Siding to be placed vertically must be purchased in lengths that will minimize horizontal joints except at wall openings:

lin. ft. of exterior wall (from floor plan openings)
– half the width of window and door openings

= x (linear feet less half the openings)
× factor in Table 6 in Appendix A

= number of panels or sheets required

Battens—The number of battens is the same as the number of boards. To calculate the number of sheets of plywood siding needed, multiply the linear footage of exterior wall by .25 (for 4-foot sheets). For each opening larger than the panel (32 square feet), deduct one sheet. If specifications call for individual vertical boards, they will require blocking or other horizontal nailers, and you must include them in the estimate.

Frieze Boards and Molding—The linear feet required is equal to the perimeter of the roof plus 10 percent of the perimeter for waste. If frieze boards or moldings are not required at gable ends, omit them in figuring the perimeter.

Fascia and Molding—Fascia and molding is found in the cornice detail of the wall sections (Figure 4-18.) The total linear footage of fascia is the roof perimeter plus 10 percent for waste. The gable end measurement equals rafter length.

Fascia Backing—The total eave length is the linear feet needed.

Soffit—In planning the cutting layout for plywood for soffit, you must allow for waste. A 4 × 8-foot sheet provides the following:

4 pieces, 1′ × 8′
2 pieces, 2′ × 8′
1 piece, 3′ × 8′
1 piece, 4′ × 8′

Figure 4-18. Sample Cornice

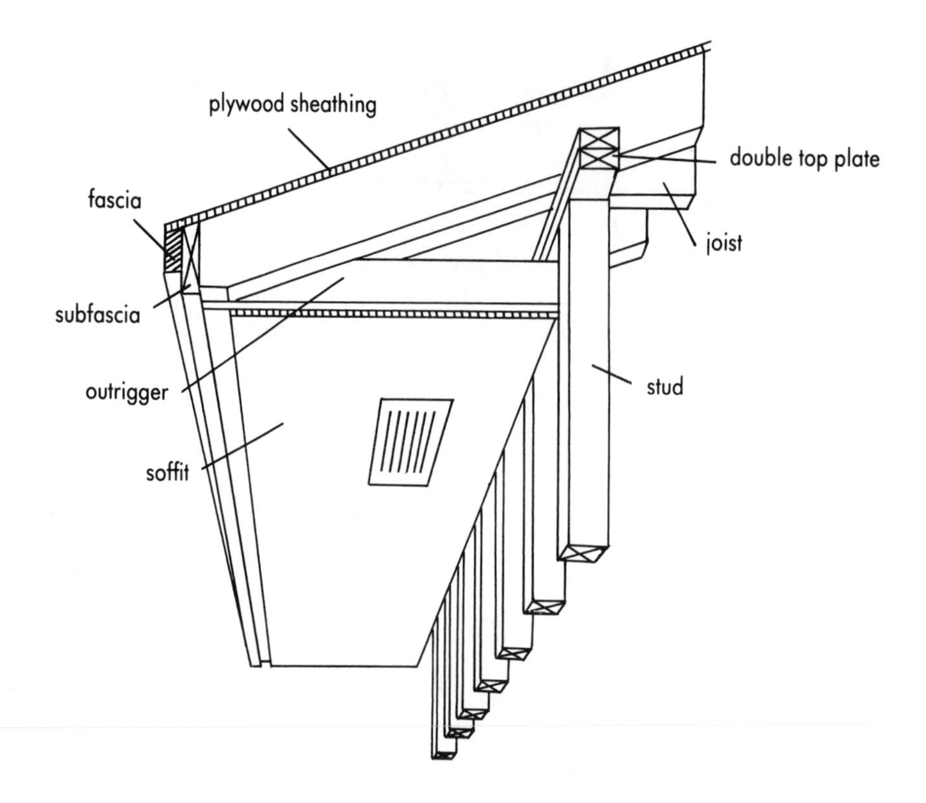

The number of sheets required equals the eave length divided by the number of pieces or sheets.

Lookouts and Dormers—(See framing detail.)

Door-and-Window Schedules

The schedule of openings includes doors and windows, along with their sizes, types, finishes, and hardware. To determine the number of doors and windows required, count from the floor plan, and list the total on the takeoff.

Doors may be prehung with or without trim and casing. Aluminum windows also may require the takeoff and purchase of separate trim and casing. To determine the quantities of casing material needed for doors, multiply the height by 2, add the width, and add 10 percent of the total for waste. If both sides of the door are to be trimmed, multiply this measurement by 2. For window trim, multiply the height and width of each window by 2 and add 10 percent for waste.

Roofing

During your review of the plans, make note of framing members including the lengths indicated on other drawings, such as the elevations, that may not appear on the principal roof-framing drawings. Measuring from and checking off on the plans provides the quantity takeoff of the roof-framing materials and the work items. To determine the area of roofing material required, you should multiply the flat area of the roof by the rafter or roof-area factor (Table 7, Appendix A). The area is usually converted to squares for pricing (and 1 square equals 100 square feet).

Method for Determining the Quantity and the Pricing Unit

Shingles—The various types and grades of shingles, as well as the method of placement, determine the number of bundles required per 100 square feet. Allow an extra bundle of shingles for each 120 linear feet of ridge and hip and each 240 linear feet of eave (Table 9, Appendix A).

Roofing Felt—This material is measured in squares of roofing. To obtain that figure, multiply the flat area of the roof by the rafter or roof-area factor in Table 7, Appendix A. Felt is bought in rolls, and a roll of 15-pound felt covers 400 square feet of sloped area. Convert the squares measured into rolls. Usually 30-pound felt is required for wood shakes. If the exposure for the shakes is 10 inches, a roll of 30-pound felt will cover 1.1 squares because of the lap required even though the roll equals 2 squares.

Roof Flashing—Metal flashing may be required along the valley(s) and as a drip edge along the eave of the roof. The ridge cap may also be metal, and flashing is needed around all projections through the roof such as vent pipes, flues, and chimneys. Rolls of metal flashing material 50 feet long are available in widths of 14 and 20 inches.

Ridge Roll and Edging—If it is required, ridge roll and edging comes in a standard 10-foot length.

Built-Up or Single-Ply Roofing—The design of a built-up or single-ply roof depends on the slope required. These types of roofs are also computed in squares. For figuring the felt for built-up roofing or the membrane requirement for single-ply roofing, you would calculate the number of squares of the roof and the number of plys (in the case of a built-up roof) to determine the layers of felt (5 ply = 5 layers). The number of squares multiplied by the number of layers or plys equals the total number of squares needed.

Gutters and Downspouts—Gable roofs need gutters on the two edges with eaves, while hip roofs need them on all four edges. The total length of the eaves that receive gutters determines the linear feet of gutter required. Count the number of downspouts from the elevations and measure their lengths.

Plumbing

Most builders subcontract the plumbing; heating, ventilation, and air-conditioning; and the electrical portions of their work. If any of these items are subcontracted you have three ways to price them:

- Guess what the subcontractors will bid.
- Get a firm quote.
- Work out some arrangement with the subcontractors for you to price the work based on unit prices.

Plumbing may be priced at so much for each type of fixture with or without the piping, or for a standard two-and-a-half-bath house, you may arrange a set price plus so much for additional fixtures. This sort of arrangement saves estimating time. If the sewer and water tie-in is not included in the plumber's base bid, you should be sure to determine that cost. You also should check that the sewer can be reached by gravity flow.

Electrical

The main difference between this subcontract and plumbing or heating, ventilation, and air-conditioning is that the builder usually furnishes some material for electrical whereas for the other two trades, the subcontractors usually furnish all the materials. In most areas builder's customarily furnish the lighting fixtures for single-family residential construction. These fixtures may be handled as an allowance. However, if the fixtures are specified, the best practice is to price them individually because of the wide variation in fixture prices.

Most electrical subcontractors figure their prices based on so much per drop (for such items as outlets, switches, and fixtures) plus so much per amp to cover the cost of the service. You should check the details regarding the kind of service required. You may incur additional costs depending (a) on whether the service is overhead or underground and (b) on other local practices or requirements. If you can work out some kind of arrangement with your electrical subcontractor for a unit price method of estimating the work, again you will save valuable time in getting the final price.

Heating, Ventilation, and Air-Conditioning

Many plans and specifications fail to show any detail regarding the heating, ventilation, and air-conditioning. The capacity of the units commonly are not specified. But for residential construction this situation usually does not work a hardship. The heat-loss and heat-gain calculations are relatively easy, and many power companies perform them for free. Also, the ductwork arrangements are fairly simple. Here, again, the most complete method of pricing the work is to get subcontract bids. However, this degree of accuracy is purchased at the expense of time.

Some mechanical subcontractors will allow you to price the work on some unit-price basis such as so much per ton plus so much for such items as humidifiers and electrostatic cleaners.

Insulation

Wood-framed floors, walls, ceilings, and roofs are insulated with either batt, blanket, or blown insulation. The area of wall to be insulated is the net area: the gross area minus the openings. The area for ceilings is the gross area. Check the specifications for special insulation for such places as along the sill and around electrical and plumbing holes.

Interior Wallboard

To estimate the amount of interior wallboard use the following calculations:

lin. ft. interior wall × height × 2 sides
− sq. ft. interior openings (6 ft. wide or more)
= interior sq. ft. − openings

lin. ft. exterior walls × height
− sq. ft. interior openings (6 ft. wide or more)
= exterior sq. ft. − openings

interior sq. ft. − openings
+ exterior sq. ft. − openings
= sq. ft. of wallboard needed

Divide the area/board to determine the number sheets required.

Ceilings—The area of the ceiling to be covered usually is equal to the area of the house. For areas with sloped ceilings, you need to increase the amount according to the slope required. For special acoustical ceilings, you may need to compute the area of each room.

Gypsum Trim—The number of exterior corners equals the number of pieces of corner trim required. Measuring edges of windows and other openings will give you the linear feet that will require this *U*-shaped trim. Usually drywall finishers furnish this trim along with the compound, tape, and nails (or screws), but you should be sure you know the local customs.

Tape and Float—For every 1,000 square feet of gypsum board for walls and ceiling, allow 375 linear feet of tape and 40 pounds of compound.

Interior Trim

Method for Determining the Quantity and the Pricing Unit

Interior Paneling—To determine the number of sheets of paneling required, divide the length of the walls to be paneled by 4 (for 4-foot sheets).

Baseboard—For linear feet needed, multiply the linear feet of interior wall by 2 and add the linear feet of exterior wall. This computation applies as well for crown molding. Add 5 percent for waste for the crown; the door openings usually take care of the waste allowance for the base.

Furring—Filling in an area over cabinets or dropping a ceiling area to conceal ductwork may require furring. Furring strips also may be installed over a surface to level it.

Strips—Strips are often of 1 × 4- or 1 × 3-inch material spaced 24 inches on center to attach thin wall panels, such as plywood.

Shelving—Measure each shelf and add 10 percent for waste.

Closet Rods—Measure linear footage of rods required from the closets on the floor plan.

Miscellaneous Millwork—Check the plans for such items as mantels for fireplaces, built-in bookcases, and cedar-lined closet shelving to make sure the millwork list is complete.

Dryer Vent—Check whether extra hose is needed.

Finish Hardware—Surface-mounted items are usually described in the specifications.

Hinges—For swaged, unswaged, mortise, half-mortise, surface-mounted, and half-surface hinges, calculate 1 pair per interior door and 1½ pairs per exterior door.

Locks—Count each privacy lock, passage lock, bifold pull, and key lock.

Coat Hooks—Count the number required.

Mirrors—Count the number and note the size required.

Bathroom Accessories—Towel bars, toilet paper dispensers, and medicine cabinets are not always shown on the plan. The number based on number of bathrooms. Check for extra towel bars.

Door Stops—Count number required.

Shelving Brackets—Count number required, 2 or 3 per shelf, depending on the size of the shelf.

Labor—Usually labor is based on square footage, but sometimes it is on a per-piece basis, such as so much per door or lineal foot of base or crown.

Painting and Wallcoverings

The room-finish schedule includes the material and finish for the baseboard, walls, and ceilings (plus height). To determine the amount of paint, stain, or sealer required, divide the square footage of area to be covered by a factor listed in Table 15, Appendix A. Because of variation in the materials, consult the supplier or manufacturer about the product before attempting to develop an accurate estimate. Many manufacturers publish their own coverage information. When paint or stain is sprayed, you will lose as much as 30 percent coverage to spray drift.

Wallcoverings are available in specific widths, so the best procedure is to round the width of the wall off to the next highest multiple of product width. In the case of repeating pattern, the height of the wall must be rounded off to the next highest multiple of the pattern height. No deductions are made for openings in the wall because the unused, cut material will most likely, be an unsatisfactory size or pattern for other areas.

Painted and stained trim is measured and estimated in linear feet. All trim up to 1 foot wide is calculated as 1 square foot per linear foot. Paint quantities used for trim are figured separately from paint quantities used for walls (Table 15, Appendix A.)

Finish Flooring

Material of a specific color, including carpet, manufactured tile, wallpaper, and vinyl, must be ordered in sufficient quantity to allow for waste. The color of the reordered material may vary slightly and may not be acceptable to the customer. Follow the suggestions listed in the paragraphs that follow to order sufficient material without oversupplying the job.

Method for Determining the Quantity and the Pricing Unit

Carpet and Roll Vinyl—The standard width for carpet is 12 feet, and 6 and 12 feet are the standard widths for roll vinyl flooring. Because waste pieces can be used in small rooms, closets, and halls, you should take the time necessary to determine the minimum amount of carpet required because the carpet supplier has little incentive to do so. All carpet must run in the same direction unless it is separated from the other carpet. The amount of carpet padding needed is the same as for the carpet.

Tile floors—To allow for waste when calculating the tile needed for individual tile floors, you should increase each room dimension to the next multiple of the tile size (9, 12, or 15 inches).

Adhesives—Check the manufacturer's information on square footage of coverage to determine the gallons of adhesive required. Also check against similar previous jobs to be sure your use is at a similar rate.

Wood Flooring—Measure each room individually and increase the width dimension to the next highest whole multiple of the flooring width. Multiply by the length to determine the square footage. Convert the square feet to board feet by multiplying by the appropriate factor in Table 6, Appendix A. Wood-flooring waste may be as high as 20 percent on small quantities and as low as 4 percent on large quantities. Check these details with the suppliers or manufacturers.

Cabinets

In your calculations and your takeoff, separate base cabinets from those that are wall hung, as well as kitchen units from bathroom units. Measure the linear footage of cabinets and built-in units. Check the details and specifications of the cabinets, if they are available. Many plans specify cabinets as standard modular units. Be sure to include spacers and end panels if they are required.

Method for Determining the Quantity and the Pricing Unit

Countertops—Measure in linear feet and count the number of finished ends.

Backsplash—Multiply the length of the area between the countertop and the bottom of the wall-hung unit by the width of the material as sold (24 or 26 inches) to determine the square footage required.

Exterior

Method for Determining the Quantity and the Pricing Unit

Final grade—Estimate the time required for whatever equipment you use for the finish grading.

Landscape—Determine the number and size of shrubs and trees. Measure the square footage of grass required. Count, measure dimensions, and calculate as needed for the details of any other landscape features such as timbers, pools, gravel, or pavers.

Walks—Measure the square footage required and any formwork necessary.

Patios—Measure the square footage required and any formwork necessary.

Drive—Measure the square footage required and any formwork necessary.

Cleaning

Method for Determining the Quantity and the Pricing Unit

Inside—Usually inside cleaning is based on square footage.

Outside—This cleaning includes hauling away all the trash or otherwise disposing of it.

Close Out and Markup

Method for Determining the Quantity and the Pricing Unit

Sales Tax—You can handle sales tax two ways: One way is to multiply each item by the proper percentage as you take it off. The other way is to list the materials in a separate column and multiply the total materials cost by the tax rate.

Worker's Compensation—You must pay Worker's Compensation Insurance premiums based on your direct labor. Again you have two choices: Some estimators add the Worker's Compensation to each applicable category. This method allows you to figure the appropriate rate based upon the craft needed for that category. The other way is to total all the direct labor for the job and add an average Worker's Compensation rate. Many subcontractors have their own insurance. If a particular subcontractor does not, the builder customarily withholds the premium amount.

Direct House Cost—The total of all materials, labor, subcontracts, equipment, sales tax, and Worker's Compensation is the direct house cost. Many builders like to determine the cost per square foot on this basis for further comparisons.

Land—List the cost of the land.

General Overhead—General overhead includes all expenses that you cannot or do not customarily charge to the jobs. These expenses include such items as office rent, electricity, heat, office supplies, furniture, legal expenses, telephone, salaries of the office employees, and salaries of the executives.

In figuring your general-overhead expenses you should pay yourself and any other executives a reasonable salary, one equal to what you could get if you worked for someone else doing what you are currently doing. This salary (or salaries) is a cost of doing business. Most construction companies operate at a general overhead rate of 3 to 10 percent. If you can keep your general overhead to about 5 or 6 percent, by all accounts you are running an efficient operation. If you are successful otherwise, do not be alarmed if your general overhead runs higher than 10 percent. However, you should be able to handle more work than you are presently doing without increasing your expenses.

In order to know how much general overhead to charge, simply divide your annual home-office expenses by your annual volume.

Profit—The best way to look at profit is that this is the amount of money that is left over after all the bills, including your own salary, have been paid. It is the amount of money left over to reinvest in the company, give out as bonuses, or use otherwise. In other words, if you make zero profit for a year but do not go below that, you have paid all the bills plus given yourself reasonable compensation. Knowing how much profit to add is a matter of time and place. At some times, you can get more than at other times, and in some places you can get more than in others.

Example—In Figure 4-19 you will find a sample estimate for the house whose plans and specifications are given in Chapter 3, "The Quantity Takeoff Process." In reviewing the takeoff, if you note that some of the prices seem out of line for your area, focus on the format rather than the prices.

The materials are kept separate for ease in calculating sales tax and ordering. The labor is broken out. On this job the only labor is carpentry and the superintendent. Each category is broken out in a manner that the authors have found most useful in their experiences. You may need to expand or combine categories. When making your own categories, keep in mind that the estimate is the basis for comparing the costs to the budget.

Figure 4-19. Sample Estimate

PRACTICAL™ FORM 510

ESTIMATE WORK SHEET

JOB Lot 2, Block C, Wood Chase 5/12 P. 1

Description		Materials	Labor	Subcontract & Equipment	Total
Job Overhead					
Plans	Lump Sum			300	
Appraisal	Lump Sum			200	
Construction Loan Closing	Loan Origination Fee, Title Exam, Deed Prep, Record Deed, Atty Fee			220	
Permit	$15 Plan Review + $3/1000 Construction Cost			195	
Inspections	3 at $30			90	
Tap Fees	Water Tap Fee-$255, Sewer Tap Fee-$600, Temp Power Hookup-$15			870	
Temp Utilities	From Records of Similar Jobs			80	
Insurance	Bldr's Risk-$100, Liability-$35, Completed Operations-$50			185	
Interest	1st month $20,000 x 1% + 2nd Month 40,000 x 1% + 3rd month 60,000 x 1% (12% annual)			1,200	
Lot Closing	Record Deed, Title Exam, Closing Fee, Taxes			200	
Sales Commission	$90,000 x 6%			5,400	
Sales Closing Costs	2nd Appraisal, Photos, Closing Fee, Document Prep, Origination Fee			1,350	
Superintendent	$40,000 ÷ 12 houses per year		3333		
			3333	10,290	13,623
Site Work					
Grading	$300 Mobilization 8 hr at $60/hr			780	
				780	780
Footings & Slabs					
Batter Boards	16 / 2x4x8 @ 1.65	26			
Formwork	Formboards 13/2x8x14 @ 6.34	82			

MFD. IN U.S.A. FRANK R. WALKER CO., PUBLISHERS, LISLE, ILLINOIS

Figure 4-19. Estimate (continued)

PRACTICAL™ FORM 510

ESTIMATE WORK SHEET

JOB Lot 2, Block C, Wood Chase S/D — P. 2

Description		Materials	Labor	Subcontract & Equipment	Total
Formboards	Brickledge 3/ 2x6x12 @ 4.16	13			
	Stakes (100) 25/ 2x4x8 @ 1.65	41			
	Braces (50) 12/2x4x8 @ 1.65	20			
	Garage Forms 2/2x4x12 @ 2.64	5			
Dowels	Smooth Dowels 14/ 5/8"ϕ x 2'-0" @ 0.40	6			
Rebars	23 @ 2.65	61			
Anchor Bolts	30 each @ 0.60	18			
Wire Mesh	1702/700 = 2.5 Say 3 rolls @ 42	126			
Gravel	1702/81 = 21 cu yd x 1.6 = 33 Tons @ $3.50 + $90 delivery	206			
Vapor Barrier	1800 sq ft	42			
Concrete	Ext Ftgs 175/(1.33x1.67)/27 = 15 Interior Ftgs 26/54 = 0.5 SLAB 1702/81 = 21 Total: 15 + 0.5 + 21 = 36.5 Say 37yd @ 55	2035			
Form & Pour	1702 sq ft @ 0.70			1191	
		2,681		1,191	3,872
Masonary					
Brick Veneer	Walls - 27x2.5x7 bricks/sq ft = 473 Total = 2601 Fireplace - 19x16x7 bricks/sq ft = 2128 5 cubes x 550 = 2750 @ $200	550		575	
Sand	at 1 yd per 1000 brick - 3 cu yd @ $20	60			
Mortar Mix	at 140 bricks per sack 2750/140 = 20 sacks @ $4.50	90			
Wall Ties	one box	12			
Fireplace	one prefab unit with chimney	689			
		1401		575	1,976

MFD IN U.S.A. FRANK R. WALKER CO., PUBLISHERS, LISLE, ILLINOIS

Figure 4-19. Estimate (continued)

PRACTICAL™ FORM 510

ESTIMATE WORK SHEET

JOB Lot 2, Block C, Wood Chase S/D

P. 3

Description			Materials	Labor	Subcontract & Equipment	Total
Framing						
Pressure Treated Sill	2X8 PT 150 lf	@ 54¢	81			
	2x4 PT 250 lf	@ 27¢	68			
Plates and Blocking	2x4 1,200 lf	@ 21¢	252			
Studs	2x4 precut 560 each	@ 1.55	868			
Headers	2x12 180 lf	@ 1.03	185			
Beams	2/2x12x18	@ 16.15	32			
Wall Sheathing	1/2"x4x8 31 sheets	@ 2.45	76			
Plywood Corners	1/2"x 4x8 CDX 16 sheets	@ 6.40	102			
Trusses	10 @ 20' common	@ 40	400			
	5 @ 20'-4" sissors	@ 40	200			
	3 @ 14'-10" sissors	@ 28	84			
	2 @ 10'-4" sissors	@ 21	42			
	18 @ 19'-9" stub	@ 40	720			
	6 @ 13'-4" stub	@ 27	162			
	1 @ 14'-10" gable	@ 30	30			
	1 @ 9'-8" gable	@ 20	20			
	1 @ 19'-9" gable	@ 40	40			
	1 @ 13'-4" gable	@ 26	26			
Rafters	17 / 2x6 x16	@ 5.25	89			
Bracing	2x4 300 lf	@ 23¢	69			

MFD IN U.S.A.

FRANK R. WALKER CO. PUBLISHERS, LISLE, ILLINOIS

Figure 4-19. Estimate (continued)

PRACTICAL™ FORM 510

ESTIMATE WORK SHEET

JOB Lot 2, Block C, Wood Chase S/D

P. 4

Description			Materials	Labor	Subcontract & Equipment	Total
Subfacia	2x8 100 lf	@ 44¢	44			
Rake Framing	2x6 250 lf	@ 30¢	75			
Lookouts & Ledgers	2x4 120 lf	@ 23¢	28			
Roof Sheathing	1/2" x 4 x 8 CDX 66 sheets	@ 8.55	564			
Entry Post	1 / 6x6x8 redwood	@ 28.80	29			
Fence Posts	4 / 4x4x8 PT	@ 4.16	17			
Fence Beams	6/2x4x8 PT	@ 2.08	13			
Beveled Siding	4000 lf 1x6 redwood (beveled)	@ 55¢	2200			
Corner Boards	180 lf 1x6 redwood	@ 60¢	108			
Window & Door Trim	200 lf 1x6 redwood	@ 60¢	120			
Frieze	90 lf 1x6 redwood	@ 60¢	54			
Facia	250 lf 1x8 redwood	@ 80¢	200			
Soffit	1/2" x 4x8 AC Plywood 5 sheets	@ 10.50	53			
Soffit Vent	90 lf	@ 50¢	45			
Louver Vents	4 each	@ 51.00	204			
Exterior Doors	$1/3^0 \times 6^8$, $1/2^8 \times 6^8$		250			
Sliding Glass Door	$1/6^0 \times 6^8$		315			
Locksets	2 each	@ 26.00	52			
Plywood Clips	1 box		13			
Felt	6 rolls	@ 9.00	54			
Fence Boards	44 / 1x6x6	@ 2.40	106			

MFD. IN U.S.A.

FRANK R. WALKER CO., PUBLISHERS, LISLE, ILLINOIS

Figure 4-19. Estimate (continued)

PRACTICAL™ FORM 510

ESTIMATE WORK SHEET

JOB Lot 2, Block C,

P. 5

Description		Materials	Labor	Subcontract & Equipment	Total
Windows	2/ 28 x 32/32 single	264			
	3/ 28 x 28/28 single	374			
	1/ 28 x 16/16 single	133			
	1/ 28 x 32/32 double	157			
	2/ 28 x 28/28 double	276			
Nails	lump sum	200			
Garage Door	1 / 16° x 17°	785			
Labor	1702 x $1.80 / ft²		3,064		
		10,279	3,064		13,343
Roofing					
Roof Edge	222 lf @ 15¢	33			
Valley Flashing	50 lf @ 64¢	32			
Shingles	21 SQS @ $24 matls & $11 sub	504		231	
Nails		36			
		605		231	836
Plumbing					
House				2,600	
Tempered Glass Door		220		100	
Sewer & Water Tie-in				360	
		220		3,060	3,280

MFD IN U.S.A.

FRANK R. WALKER CO., PUBLISHERS, LISLE, ILLINOIS

Figure 4-19. Estimate (continued)

PRACTICAL™ FORM 510

ESTIMATE WORK SHEET

JOB Lot 2, Block C, Wood Chase S/D

P. 6

Description		Materials	Labor	Subcontract & Equipment	Total
Electrical					
Wiring	Lump Sum			1,600	
Light Fixtures	Allowance	1,000			
		1,000		1,600	2,600
HVAC					
House	2½ T			2015	
				2,015	2,015
Insulation					
Attic	R-38 12" Batts 1000 sq ft @ 66¢			660	
Walls	R-13 3⅝" Batts 1920 sq ft @ 32¢			615	
Vaulted Clg	R-30 9" Batts 320 sq ft @ 52¢, Rigid 320 sq ft @ 18¢			224	
				1,498	1,498
Drywall					
Garage	5/8 Firecode 930 sq ft 20 bds @ 8.65	173			
House	5/8 Regular 4970 sq ft 104 bds @ 7.78	809			
Labor	Hang and Finish 124 bds @ 10.56			1309	
		982		1,309	2,291
Interior Trim					
Base	440 lf @ 32¢	141			
Casing	530 lf @ 27¢	143			
Closet Rods	26 lf @ 37¢	10			

MFD. IN U.S.A. FRANK R. WALKER CO., PUBLISHERS, LISLE, ILLINOIS

Figure 4-19. Estimate (continued)

PRACTICAL™ FORM 510

ESTIMATE WORK SHEET

JOB Lot 2, Block C, Wood Chase S/D

P. 7

Description		Materials	Labor	Subcontract & Equipment	Total
Shelving	40 lf 1x12 White Pine @ 76¢	30			
Hardware	5 Privacy locks, 7 Bifold Pulls, 9 Bumpers	167			
Mantle	1x2x8 oak, 1x4x18 oak, 1x6x12 oak, 3" oak crown - 14 lf, oak bead Mldg - 24 lf	60			
Bath Accessories	Paper Holders, Soap Dishes, Toothbrush Holders, Towel Bars / 2 ea	125			
Mirrors	2 / 4'x4' = 32 sq ft @ $3	96			
Interior Doors	7 wing, 7 Bifold @ 65.00 Avg	910			
Nails	Lump Sum	150			
Shower Rod	2 each @ $11	22			
Dryer Vent	Vent plus 20' hose	20			
Labor	1274 sq ft @ $1.10		1,401		
		1,874	1,401		3,275
Paint & Wallcover					
Exterior	4720 sq ft @ 20¢			944	
Interior	1273 sq ft @ 50¢			653*	
Wallpaper	18 rolls @ $20.00			360	
				1,957	1,957
Floor Covering					
Carpet	100 sq yd @ $15.00			1,500	
Vinyl	350 sq ft @ $2.00			700	
				2,200	2,200

*The calculation error in the takeoff of this item was left as an example of how easily a mistake can occur in a takeoff done by hand. See also "Accuracy" in Chapter 7, "Computerized Estimating."

PRACTICAL™ FORM 510

ESTIMATE WORK SHEET

JOB Lot 2, Block c, Wood Chase s/D — P. 8

Description		Materials	Labor	Subcontract & Equipment	Total
Cabinets					
Kit & Bath		1508	65		
Appliances					
Range		625			
Refrigerator		850			
Dishwasher		290			
		1,765			1,765
Exterior					
Final Grade	8hr @ $45			360	
Landscape	Shrubs & Trees - 18 @ $15 in place, Grass - 11,170 sqft @ 7¢			1,089	
Walks	5/ 4'x4' x 1.50	120			
Patio	130 sqft @ 2.00			260	
Drive	16 x 60 = 960 sqft 960/81 = 12 cu yd	660			
		780		1,997	2,777
Clean					
Inside	1670 x 15¢			250	
Outside	Lump sum - 3 trips			600	
				850	850
Subtotal		23,095	7,863	29,553	60,511

MFD. IN U.S.A. FRANK R. WALKER CO. PUBLISHERS, LISLE, ILLINOIS

Figure 4-19. Estimate (continued)

PRACTICAL™ FORM 510

ESTIMATE WORK SHEET

JOB Lot 2, Block C, Wood Chase S/D

P. 9

Description		Materials	Labor	Subcontract & Equipment	Total
Subtotal		23,095	7,863	29,553	60,511
	Tax 23,095 x 0.07				1,617
	Worker's Comp 7,863 @ $5.70/100				448
Direct House Cost					62,576
Land					15,700
Subtotal					78,280
	General Overhead @ 5%				3,914
	Profit @ 10%				7828
Sales Price					$90,018

MFD. IN U.S.A. FRANK R. WALKER CO. PUBLISHERS LISLE, ILLINOIS

Chapter 5

Accuracy in Estimating

Inaccuracy in an estimate can come from several different sources. One source of error comes from not going into enough detail. However, not going into detail is not always bad because of the relationship between accuracy and time: given enough time an experienced estimator can be quite accurate. You must use careful judgement when deciding how much detail to give to a certain category or item. Generally, you should try to work out a fast, reasonable basis for estimating on low cost items.

Example—You could count all the nails for a job allowing for waste and probably come up with a reasonable number. But why waste time doing that when you can look back at similar jobs and see that you typically spend from $175 to $200 for nails?

Based on the experience of many builders, if you use the degree of detail shown in the example in the Chapter 4, "Preparing a Complete Estimate," you can expect to be within 3 to 5 percent given no unforseen contingencies.

A second source of error results from blunders. Perhaps the most costly blunder of all is to leave something out. Guessing the cost of an item at half of what it really costs is better than leaving it out and being off by twice as much. Also, if you rush through the arithmetic part of the estimate, the probability of error increases.

The best approach to eliminating errors in your estimate is to concentrate on each entry and calculation. This concentration requires a sufficient amount of uninterrupted time to complete the work.

Clerical errors are usually caused by one of the following:

- Using the wrong units or unit cost
- Misplacing or transposing numbers
- Improper placement of decimals
- Addition or multiplication mistakes

Example—An incorrect calculation might occur as follows:

Quantity	Unit Cost Material	Cost Concrete Footings
270 cu. ft.	$48.50 per cu. yd.	$13,095

In this case, you must convert the cubic footage to cubic yards (270 cu. ft. ÷ 27 cu. ft. = 10 cu. yd.) before the cost extension provides the correct cost estimate:

$$10 \text{ cu.yd.} \times \$48.50 = \$485 \text{ not } \$13{,}095$$

Although the use of a paper tape calculator allows you to check entries against the information listed on the takeoff and cost-extension sheet, it does not alert you to any errors in quantities or unit costs listed. Conducting a thorough review of the quantities and unit costs can resolve any apparent discrepancies.

To reduce addition errors, total the combined labor, material, and subcontract costs of each item and compare this to the sum of the labor column, the material column, and the equipment column. The estimate sheet is complete and correct when the total page cost—the last amount in the far-right column—equals the sum of the line item costs and the sums of the individual cost-category columns. (Figure 5-1 illustrates this point.)

Pricing Units

Unit costs used to complete the cost estimate are based on the quantities determined in the takeoff. You need to be familiar with and use the standard units of measurement. If you know in advance the pricing unit that will be used, you can complete the quantity determination and takeoff using those units.

Example—Roofing labor and material is usually bought by squares:

$$\frac{\text{flat area (sq. ft.)} \times \text{rafter factor}}{100} = \text{squares required}$$

Material quotes are usually based on the unit of measure used to invoice the material. Therefore, a material, such as concrete, that is sold in cubic yards is priced in cubic yards. Direct labor is bought on a workhour basis. Converting workhour cost to material units is easier than converting material units to workhour cost. Therefore, many labor cost estimates for work items use the same pricing units as materials. This practice requires that you compute labor unit cost by measuring the labor cost for an item of work and dividing this measurement by the units of material used.

Unit-cost labor quotes may or may not be in the same pricing units as the material. Concrete placement is often quoted by the square footage of the slab, while rough carpentry is often quoted by the square footage of total area.

Labor Cost Estimates

Direct labor estimates are harder to estimate accurately than any other cost category. The labor cost estimate must reflect the anticipated cost considering reasonable worker productivity, the site, and weather for the construction period. Your historical cost record is the usual source for labor unit costs.

Your checklist for labor items should be similar for each house you build that is of the same general size and complexity. Likewise, the labor productivity and wage rates calculated for each unit cost are similar or the same for work items.

Labor unit costs vary when wage rates change or when job cost records indicate worse or better productivity than previously recorded

Figure 5-1. Sample Foundation Cost Estimate

FOUNDATION COST ESTIMATE

Item	Quantity	Labor Unit Cost	Material Unit Cost	Subcontract Cost	Labor Cost	Material Cost	Total
Excavate Footing	Lump Sum	—	—	$ 100 –	—	—	100 –
Footing Concrete	10 CY	10 –	52 50	—	100 –	525 –	625 –
CMU Foundation							
8x8x16 Block	100 EA	—	.67	60 –	—	67 –	127 –
Header Block	116 EA	—	.88	70 –	—	102 –	172 –
Mortar Mix	8 SK	—	2 95	—	—	24 –	24 –
Sand	1 YD	—	25 –	—	—	25 –	25 –
Soil Poison	Lump Sum	—	—	75 –	—	—	75 –
Slab Gravel							
1280 SFT / 16 YDS	24 TON	—	3 50	—	—	84 –	84 –
Hauling	2 LOADS	—	—	60 –	—	—	60 –
Spreading	Lump Sum	—	—	75 –	—	—	75 –
WWF							
1280 SFT	2 ROLLS	w/SLAB	40 –	—	—	80 –	80 –
Vapor Barrier							
1280 SFT	1 ROLL	w/SLAB	30 –	—	—	30 –	30 –
Edge Insulation							
150 LFT	300 SFT	w/SLAB	.50	—	—	150 –	150 –
Slab Concrete							
(Incl. Ext.)	27 CY	w/SLAB	52 50	—	—	1419 –	1419 –
2160 SFT		.30	—	—	648 –	—	648 –
Page Total				$ 440 –	$ 748 –	$ 2506 –	$ 3694 –

and estimated. Therefore, when pricing, you should use careful judgment in applying the quantity takeoff information in light of historical labor cost-estimating information. Figure 5-2 provides a sample of a cost record for footings for completed houses.

The record in Figure 5-2 shows a range of footing costs from $20 per cubic yard to $50 per cubic yard. If poor site or weather conditions are anticipated, you need to use a higher unit cost to estimate the cost for the work item, so you have a contingency. Before totaling the labor cost you should review individual labor estimates for each item. If you know the wage rates and average crew sizes, you can check the amount of worktime that is budgeted for the labor cost estimate.

Figure 5-2. Sample Footing Costs

Item	House	Actual Cost	Quantity	Price/Unit	Remarks
Footings	Smith	$ 500	10 cu.yd.	$50/cu.yd.	Rain, hauled in wheelbarrows
	Jones	$ 500	20 cu.yd.	$25/cu.yd.	Good condition
	Doe	$1,000	50 cu.yd.	$20/cu.yd.	Large job, excellent condition
	Spec. 1	$ 275	10 cu.yd.	$27.50/cu.yd.	Average house and site
	Spec. 2	$ 319	11 cu.yd.	$29/cu.yd.	Average house and site

Example—The total cost of labor for a particular item of work is $965, based on a unit labor cost of $50 per unit for 1,930 units of work. The budgeted crewtime, at a cost of $32 per hour is as follows:

$$\frac{\$965}{\$32/\text{crewhour}} = 30 \text{ crewhours} \qquad \frac{\$965}{\$256/\text{crewday}} = 3.8 \text{ crewdays}$$

If the estimator in this case is confident the actual workup required by the plans and specifications can be completed by a $32-an-hour crew in 30 hours or 3.8 days, the labor estimate cost is satisfactory. Otherwise, either the quantity is incorrect or the estimated unit cost ($50 per unit) is inaccurate for the quantity or the quality of work required.

In building, the greatest cost variance from job to job traditionally has been in the labor cost. Because of this variance, you may decide to add a reasonable contingency amount to the labor estimate to cover unanticipated difficulties or overtime that might be required. However, because of the trend toward subcontracting most of the labor, this contingency is now being transferred largely to the subcontractor.

The best approach to estimating a contingency labor cost is to figure the contingency based on the total labor cost estimate for a category of work, rather than on each individual work item. Compare the total labor estimate (before labor burden) to average crew cost per hour, day, or week to determine the amount of time budgeted for each category and make contingency estimates. A summary of labor, with a contingency estimate using a $32-an-hour crew comparison, appears in Figure 5-3.

The contingency of $928 provides for a contingency of 29 crew hours for a 15-week project schedule (75 workdays or 600 crewhours). This contingency equals 4.8 percent of the base labor estimate: $928 ÷ 19,200 = 4.8%.

However, the direct labor cost estimate does not include the indirect cost of the labor burden. A substantial labor cost overrun will occur if the labor burden estimate is left out of the cost estimate.

Figure 5-3. Sample Labor Summary

Category	Estimate	Crew Weeks	Contingency	
			Percent	Cost
Foundation	$ 3,840	3	10%	$384
Rough carpentry	6,400	5	5	320
Exterior	1,920	1.5	5	96
Interior	5,120	4	0	
Roof	1,280	1	5	64
Walks/drives	640	0.5	10	64
	$19,200	15	4.8%	$928

Cost Extensions from Quantity Takeoff

An accurate cost estimate for a residence results from accurate cost estimates for each item or category required to complete the construction of the house. The total cost includes not only the direct costs but also the jobsite overhead and general overhead for the services provided. Of course, profit is not a cost: it is what is left over after all the costs are paid. Along with subcontract bids and material cost collection, you must complete an estimate of the direct costs from historical cost records for the portion of work that will be performed by your own crews.

This portion of the cost estimate requires that you list your crews and enter the related cost estimating information to complete the cost extensions. When your labor cost records do not include a cost measurement for a particular item, you can estimate that cost by comparing the work required with other work that your crews previously have done.

When collecting and evaluating subcontracts and costs for major material items, the estimator should set up and use a bid management system to complete this portion of the cost estimate. A system that includes a separate record of each bid including the telephone bid record (Figure 5-4) is necessary if the estimator is to complete the process with a method for estimate verification. The bids can be listed on a bid tabulation form (Figure 5-5) that is formatted to summarize, evaluate, compare, and select the low, qualified bid for each item.

Concentrating on one component of the cost estimate at a time provides the opportunity to make the best judgments and the fewest errors. It also produces the most complete and valid cost estimate.

Figure 5-4. Sample Telephone Bid Record

Job ______

Location ______

Date ______ Time ______

Firm ______

By ______

Classification of Work ______

Base Bid $______

Work Included

______

______

______

______

______

______

Exclusions and Qualifications

______

______

______

______

______

______

Alternate No. ______ Add/Subtract $______

No. ______ Add/Subtract $______

No. ______ Add/Subtract $______

By ______

Base Bid ______ (recheck)

Adjustment: Add/Subtract $______

Time ______

By ______

Amount to Bid Summary $______

Figure 5-5. Sample Subcontract Bid Form

SUBCONTRACT BID

Line		SUBCONTRACTOR	1	2	3 BASE BID	4	5 QUALIFICATIONS	6
1								
2		ELECTRICAL						
3								
4		WATTS ELECTRICAL CONTRACTOR						
5								
6		JOHNSON ELECTRIC COMPANY						
7								
8		MMR						
9								
10	(EXTRA)							
11								
12								
13		PLUMBERS						
14								
15		J&B PLUMBING						
16								
17		HARILSON PLUMBING, INC.						
18								
19		CITY PLUMBING CO.						
20								
21	(EXTRA)							
22								
23		HVAC						
24								
25		AA AIR						
26								
27		J&B PLUMBING						
28								
29		ROCKETT HEATING AND AIR						
30								
31		TRI-STATE HEATING						
32								
33	(EXTRA)							
34								
35								
36								
37								
38								
39								
40								

However, if you have a prearranged system with your subcontractors to price their work on some unit-price basis such as so much a square foot, fixture, or ton, this portion of the estimate goes much faster.

Material Cost Estimates

The cost extensions for material cost estimates are categorized as allowances, lump-sum quotes, lump-sum prices, unit-cost quotes, and unit-cost prices.

The lump-sum quote materials need only to be listed in the takeoff. A quantity determination is not required. Most estimators identify such quantities as LS (lump sum).

You should recognize and understand the difference between a lump-sum quote or price and a unit-cost quote or price. The vendor's quote or price, while indicating a total amount for a category of materials, may actually be qualified or limited to furnishing a specific quantity of each item. This practice is typical when the vendor's quote or price requires the vendor to prepare a quantity takeoff, as in the case of framing lumber.

This qualification of a quote or price accounts for differences between the quantity of material used and that listed in the vendor's takeoff. Therefore, you should use a lump-sum quote or price as a unit-cost quote or price and request a copy of the vendor's takeoff. If you compare your independent takeoff to the vendor's listing, you can evaluate and validate the lump-sum quote or price.

The checklist in Chapter 3, "The Quantity Takeoff Process," lists standard units for purchase, although some vendors may quote in different units. The quantity takeoff listing must be in more than one set of units if you are to compare the unit cost quotes or prices.

Example—One vendor may quote the following:

Random length—2″ × 4″ in $/bd. ft.
Premium length—2″ × 4″ in $/bd. ft.
Precut studs—2″ × 4″ in $/stud

While a second vendor may quote on 2″ × 4″ precut studs:

2″ × 4″ × 10′ in $/each
2″ × 4″ × 12′ in $/each
2″ × 4″ × 14′ in $/each
2″ × 4″ × 16′ in $/each

A third vendor might quote all materials in units of linear feet.

Again, you must be careful in listing prices furnished by vendors as compared to quotes furnished by vendors. If the vendor lists an item for which only current prices are known, the material cost estimate needs adjusting if price increases are anticipated before the materials will be ordered. However, material quotes are guaranteed by the vendor through the time of delivery. When they are based on accurate quantities, they contribute to an accurate cost estimate.

Lump-sum quotes are, likewise, different from lump-sum prices. If the material is priced, rather than quoted, on either a lump-sum or unit-price basis, you should consider the potential for cost escalation between the time of the estimate and the time of purchase or delivery of the material. This evaluation for potential increases in cost is best done on an item-by-item basis and reconsidered on a total project basis.

The appropriate method of establishing allowances is to count, measure, or assume a specific quantity and price, quote, or assume a specific unit cost. (Allowances are discussed in detail in Chapter 2, "The Complete Estimate.") After you complete the cost extensions, summarize the total allowance for the material category. (The sample foundation cost estimate in Figure 5-1 shows the cost extensions for the materials required for the foundation and concrete work for a house.)

Subcontract Cost Estimates

The classifications of subcontract cost estimates are as follows:

- Labor only, unit cost
- Labor only, lump sum
- Labor and material, lump sum

Because you will undoubtedly receive quotes from subcontractors that differ in format, you need to use a format that allows for comparison and evaluation of each of the individual items that will be subcontracted.

Many estimators set up subcontract summaries or tabulations (Figure 5-6) for comparing bids, while other estimators prefer to list the subcontract bids, especially the labor-only bids, in with their own labor and material cost estimates. When using the latter method, you need to separate labor cost estimates from the subcontract labor bids, unless you pay the labor burden for the subcontractors' crews.

The unit-cost subcontract items require an accurate estimate of the quantity in appropriate units for the subcontract estimate. When you use labor-only subcontractors, whether unit cost or lump sum, you must be certain a complete material takeoff and cost estimate supports the labor-only subcontract.

The lump-sum subcontract bids are easily evaluated. They require only a format that makes recording and comparing the bids easy.

Equipment Cost and Jobsite Overhead

The pricing units for most equipment and overhead items are units of time (hours, days, weeks). Therefore, you must list the quantities of those items in the appropriate time units. The unit cost for cost extension of equipment and jobsite overhead items also must be based on the same units of time.

Your own equipment probably supports more than one work item. For most pieces of equipment, a week is an appropriate unit for cost estimates. Other equipment, such as concrete tools that are used for shorter duration, should be estimated in daily units.

Many estimators list the equipment on a separate summary and estimate all of the equipment cost at one time. Other estimators prefer to include the equipment cost with the individual work items by adding a column for equipment unit cost and one for equipment cost. These columns are not included in the suggested format at the beginning of this chapter (Figure 5-1).

Figure 5-6. Subcontract Summary

SUBCONTRACT SUMMARY

SUBCONTRACTS	QUALIFICATIONS	BASE BID	ADJUSTMENT		BID USED IN ESTIMATE
ELECTRICAL					
WATTS	NO FIXTURES	$ 1600 -			
JOHNSON	NO FIXTURES	$ 1700 -	- $150 - 2PM		$ 1550 -
MMR	NO FIXTURES	$ 1850 -			
APEX	$800 FIXTURE ALL	$ 2375 -			
	NONE W/O FIXTURE	1575 -			
PLUMBING					
J&B	5' OUTSIDE BUILDING	$ 2725 -			
HARRISON	INCL. TIE IN	$ 3000 -			$ 3000 -
CITY PLUMBING	INCL. TIE IN	$ 3100 -			
TRI-STATE HEATING	5' OUTSIDE BUILDING	$ 3000 -			
HEAT A/C					
AA AIR	NONE	$ 2150 -			
J&B	NONE	$ 2050 -			$ 2050 -
ROCKETT	NONE	$ 2500 -			
TRI-STATE	NONE	$ 2250 -			
TOTAL					$ 6600 -

Contingency Costs

The estimating process must provide some evaluation of the anticipated conditions of construction because actual conditions often differ from ideal conditions, and they can strongly affect costs. Because you have to estimate a single cost, rather than a range of costs for each item of work, how can you estimate a competitive cost and still allow for the range of conditions that might occur?

One method of allowing for unexpected occurrences increases the unit cost of the labor and/or material for particular work items. But this method does not facilitate measuring the actual expenditure for cost contingencies on individual categories of work. In addition, normal actual costs might appear to be cost savings if the contingency cost is not needed for the work.

The best approach is to establish separate contingency cost estimates for the totals of work categories. You can use normal unit costs for estimating material and labor costs for individual work items and then apply contingency costs as a percentage of the total costs per category.

The need for a contingency labor cost results from the probability of reduced productivity on the job site. Difficult details, poor accessibility to the site, inclement weather, and labor shortages all affect productivity.

The normal contingency for the labor cost estimate might range from 1 to 10 percent of the labor cost. If overtime or cost overruns occur, you can account for amounts in excess of the labor cost estimate for the work item and subtract them from the contingency balance. As long as the balance is positive, an overall project labor cost overrun does not occur.

Experienced builders use a similar approach for unprotected material prices. A contingency balance for material price increases can offset any actual price increases that occur between the estimate and the material purchase.

Requirements for an Accurate Estimate

An estimate is accurate when the quantities are accurate, the costs are correct, and the process used for the estimate is comprehensive. An accurate estimate is essential to profitable homebuilding. If you are to achieve an accurate cost estimate, you must understand the validation process and consistently apply it through quantity determination and takeoff, subcontract bidding, and the cost extension process.

You must have confidence in the accuracy of the estimate you use to establish a contract amount for building a house or other structure. An accurate estimate meets the requirements listed in Figure 5-7.

One way to validate a cost estimate is to compare it to historical cost records. The more promptly you validate a cost for a particular item, the sooner the recorded cost is available for reference.

If the format of the cost estimate is consistent with the format of the cost records, you can compare the cost records to the cost estimate on an item-by-item basis. Using a system such as the NAHB Chart of Accounts[1] for both the cost estimate and the cost accounting system provides this compatibility.

Throughout construction of a house, you should validate the cost estimate by comparing the cost control information developed from

Figure 5-7. Requirements of an Accurate Estimate

- The quantity of material estimated equals or slightly exceeds the quantity of materials actually used for the building.
- The quantity of units in each unit-cost subcontract equals the estimated quantity for that contract.
- The unit prices in each unit-cost subcontract equal the costs used in the estimate.
- Each item in a lump-sum contract is bought for the amount used for the item in the estimate.
- The total labor cost estimate for each item and category of work equals or reasonably exceeds the amount of time estimated for a crew multiplied by the cost of the crew.
- All contingency costs are known and understood.
- The amount of time estimated for equipment and jobsite overhead items equals or exceeds the actual time period required.
- The actual rate for the equipment does not exceed the estimated rate.
- The labor burden and sales tax are accurate for the labor and materials required.
- The margin portion of the cost estimate equals the risks, costs of business, and profit expectations.

the cost estimate with the records (and computations) of actual cost. Cost control is effective only if it is timely. The process of comparing actual quantities and costs to estimated quantities and costs is the best method of controlling direct building cost. (See Chapter 6, "The Cost Control System.")

The quantity of work performed and quantity of material used is a primary consideration in validating the labor and material cost estimate, so you need to verify these costs as soon as possible. A variance in the actual quantity from the estimated quantity indicates either a problem in the takeoff or unanticipated job conditions.

Validating the equipment and jobsite cost is also a function of the job cost control system. You report actual time required, the cost rate, and the differences compared with the estimated amounts.

The job log identifies the reasons for the variance and allows you to adjust future estimates if necessary.

You must validate your estimates and unit-cost information on a continual basis. Better estimates result when you use validated previous cost estimates in preparing current and future cost estimates.

A cost estimate is validated, of course, upon completion of construction. This final evaluation provides a learning experience for improving future estimates.

Methods of Checking an Estimate

When you are completing the takeoff you should verify the quantities specified as you list them on the takeoff form.

Rechecking each individual computation is inefficient, but you need a comprehensive review, preferably by another person. Some building companies have different people perform the quantity takeoff and cost estimating, with the estimator reviewing the quantities determined. Multiplication, division, and addition errors are common in inaccurate cost estimates. Therefore, you should check each individual item on each page of the estimate and the summary pages in detail. Also check for transposition errors that may have occurred in transferring quantities, costs, or other numbers from one form to the next.

The most rapid method of checking quantities of materials is to review the material according to some known comparison factor that can be reasonably applied considering the size, time, and quality of work. For instance, you can compare many quantities to the total square feet, heated area, number of rooms, or number of fixtures. The material quantity divided by the comparison factor should fall within an acceptable range of predetermined board feet, linear feet, square feet, cubic yard, or other unit per square area or room.

Example—You might compare the roof material quantities to the heated area and anticipate .012 to .015 squares of roofing material per square foot for a single story roof. The sloped area of the roof always exceeds the flat area of the floor. If the quantity takeoff lists more than .015 or less than .012 squares of roofing per square foot of heated area of a single-story house, you need to double check the quantity calculations.

You can compare total board footage or linear footage of the framing lumber by size and type or by the number of studs per square foot of heated area. This ratio can be compared with quantity ratios of materials used for previously built houses. This comparison allows you to identify loss factors and potential quantity errors. You could use a similar method to verify the quantity of other materials.

Whenever possible, base the material cost estimate on firm quotes. Recheck the cost listed in the estimate with the quote and price information collected to validate the unit or lump-sum cost used in the material cost estimate. With experience you will learn to recognize the appropriate unit-cost range for materials and to know what to expect from cost extensions.

Labor cost estimates are often validated by comparing the total labor cost for a particular item to a standard quantity, such as one used to validate the material quantities. Compare the labor cost estimate per square foot or other factor to cost records on previous buildings.

Example—If your records indicate that your costs are $1.25 to $1.35 per square foot for a particular work item and if the estimate to be validated is $1.05 for each square foot, consider what might be different in the house being estimated. Unless the details of the house are substantially different from those in previous houses or the cost of labor on a crewhour basis is less than previously expected, you should check both the quantity of work and the crew productivity estimated for computing the unit cost for the labor.

Some builders check the entire labor cost estimate prior to checking the labor cost for the individual work items. This check is often done by two methods: comparison to historical cost records or a detailed, item-by-item verification. The first method uses the following formula:

$$\frac{\text{total labor cost estimate}}{\text{house square footage}} = \text{labor cost per square foot}$$

The second method uses the following formula:

$$\frac{\text{total labor cost estimate}}{\text{average crew cost per hour (day)}} = \text{budgeted crewhours (days)}$$

The historical labor cost per square foot and the actual crewhours (days) used in building a prior, similar house gives a reasonable comparison for the labor cost estimate.

Because the labor cost estimate is the principal risk factor in building, if you do a substantial amount of the labor with your own crew, you should spend a substantial amount of time and effort checking the labor cost estimate. The labor estimate validation uses the same process as the estimate breakdown for labor cost control (Chapter 6, "The Cost Control System").

You should recheck each subcontract cost used in the estimate. Collecting multiple bids for each item ensures that the item can be bought for the subcontracted cost estimated. The subcontract costs also are actually checked during the buy-out of the house (Chapter 6).

Checking the equipment and jobsite overhead costs includes reviewing and concurring on the time listed for each item. Likewise, the cost per unit of time must be consistent with the actual rate of cost that occurs. Again, your historical cost records are the references for checking the cost estimate for the equipment and for jobsite overhead.

Many errors occur when the time to estimate costs is insufficient for a complete review and validation of the cost estimate. To guard against this possibility, you must follow the steps in Figure 5-8.

Figure 5-8. Steps for Completing and Reviewing an Estimate

- Follow a process that uses time efficiently.
- Close out the takeoff and verify the quantities and units as soon as these actions are reasonable.
- Collect subcontractors' and material quotes as early as is reasonable or make arrangements for unit prices with your subcontractors and material suppliers ahead of time.
- Complete and check the labor and material cost extensions in advance of when the bid is due.
- Prepare and check the jobsite cost estimate in advance of the bid due date.
- Summarize and check the cost estimate for labor burden and sales tax immediately upon completing the totals for the labor cost estimate and the material cost estimate.
- Use a format for evaluating and summarizing subcontract bids, such as bid adjustment forms, that can be used accurately in a limited period of time.

The buy-out of the house determines a substantial portion of the cost. The portion of the cost estimate that can be promptly checked as actual cost, limits your risk in the project.

The buy-out includes the following items:

- Issuing lump-sum subcontracts for the amount used in the cost estimate (both for labor only and for material and labor).
- Issuing unit-cost subcontracts that specify the number of units (if possible) for the bid unit cost.
- Issuing purchase orders based on quotes and delivery dates for materials.

The balance of the cost control is accomplished throughout the term of the project. The estimate breakdown establishes the budget for the following items:

- Actual crew rates to be paid and crew productivity required for work items and categories of work
- Unprotected material prices
- Equipment time based on actual rates to be paid
- Overhead

Improving Estimate Accuracy and Quality

Because quantity determination and cost estimating are time consuming, many builders and estimators are interested in saving time without sacrificing the quality of the estimate. The following devices expedite the process and also enhance accuracy:

- Checklists and conversion tables
- Standard forms and formats for the takeoff and cost estimate
- Computation and waste/loss tables and factors
- Consistent estimating methods to help expedite the process

Most estimators agree that improving the accuracy and quality of the various elements of an estimate comes with experience. Valuable estimating experience is gained from a well-developed, standard method or approach applied to a substantial number of estimates.

The principal way to improve the accuracy of a cost estimate is to identify the appropriate level of detail required for the estimate items and to assure compliance at that level. You can waste a lot of time and effort producing a level of detail and accuracy in the quantity takeoff that is not justified by the cost information available.

Reviewing the cost estimate in relation to actual costs helps identify any problem in the estimate process, so carefully kept cost records are imperative for the accuracy of future estimates. If the unit costs and cost extensions for the estimate are based on historical records, the level of detail in the takeoff should match the level of detail of those records.

Using Cost Records

Historical cost records should include broad-scope items, as well as specific items of work. Validating overall labor cost estimates or material quantities and costs for segments of work requires the use of cost records on a house-by-house basis.

A cost control and accounting system provides records of actual cost by category at an appropriate level of detail. On a broad scope, the comparisons for items or categories of work can be listed as they are in Figure 5-9.

Your records are incomplete until you include in them the exact reasons for cost variances. A variance between actual cost and estimated cost results from any number of the occurrences listed in Figure 5-10.

You must compare an identified variance in the quantity of a material to the actual list of materials needed for the plan built. A variance might result from shortages in a delivery or a theft or damage on the job. All of these possibilities directly reduce the profit on a fixed-price contract. If an estimated quantity is inaccurate because of factors other than plan changes, the waste/loss factors used for the estimate are

Figure 5-9. Comparisons by Category of Actual Cost with Estimated Cost

	Total Estimated Cost	Total Actual Cost	Variance (+,−)	Number of Units	Estimated Unit Cost	Actual Unit Cost
Material Costs						
Labor Costs						
Subcontract Costs						
Equipment Costs						
Jobsite Costs						
Margin						

Figure 5-10. Reasons for Cost Variances

- Changes in house design/detail
- Site/weather conditions
- Changes in the work plan
- Changed labor conditions
- Labor quantity variance
- Labor wage-rate variance
- Labor productivity variance
- Material cost variance
- Material quantity variance
- Subcontract cost change (from estimate to actual)

inaccurate or a mistake was made in determining the quantity needed. If the actual quantity supplied exceeds a valid quantity, waste or loss was excessive. In that case, jobsite control was inadequate, and the quantity takeoff was correct.

Historical labor cost records include not only the actual cost incurred but also the conditions affecting the cost. The labor cost actually incurred for a work item along with the actual number of units installed (cost of labor per board feet of material) is recorded for each item.

You can assign costs to items only at a level as detailed as the cost control system provides. If you measure, record, and evaluate crew cost on a weekly basis, you can only develop gross cost factors, such as labor cost from slab to black-in. If you measure and record crew cost daily or hourly, you can segregate the cost of labor for specific categories of work, such as wall framing.

Some builders use labor unit cost for labor cost extensions, and their promptness in figuring the cost estimate depends on the format of their historical cost records. Other builders estimate their labor costs on a workhour or crewhour basis. The cost record for that type of estimate requires that the workhours and/or crewhours required be recorded beside the cost estimate items listed. The records are then applied with current wage rates to complete the labor cost estimate.

Subcontract cost records—especially unit costs and those lump sums that are appropriately related to a measurable unit—provide a valid reference for evaluating future subcontract bids, especially when single bids are received. You can also use the information in preparing preliminary estimates.

Each category of subcontract cost can be delineated to an appropriate level of detail to identify specific cost variances. Compare actual subcontract costs to the costs used in the estimate (Figure 5-11).

A record of actual expenditures compared to allowances for items helps in validating allowance estimates in the future. Increases or decreases in the amount and/or cost of items estimated by allowances should be recorded with a description of the circumstances resulting in the variance.

Most experienced builders use measurable quantities for cost control, and they use those units of measure for cost estimating.

Figure 5-11. Comparison of Estimated Costs to Actual Subcontract Costs

	Estimated Subcontracts	Actual Cost	Percentage Variance ± Cost	Variance Evaluation
Electrical				
HVAC				
Plumbing				
Sitework				
Concrete (labor only)				
Framing (labor only)				
Masonry (labor only)				
Roofing				
Seeding, landscape				

Example—Although materials for framing and rough carpentry are listed on an item basis, the labor may be estimated using the total area of the house.

The best and most efficient way to assure a complete and accurate listing of all costs incurred is to use standardized forms and a standard estimating format. If you use a similar purchasing method for each house, the estimating format should comply with the standard cost control format for accurate comparison and evaluation.

Builders who use the same plans repetitively can prepare a complete, verified list of materials and labor quantities with cost extensions for the house and reuse it each time. During the process of determining quantities, you can concentrate on the site work and the foundations required.

The cost records for future estimates are the final product of the cost control. Considering the long-term usefulness of cost records, this step is perhaps the most important one because the accuracy of future cost estimates based on the cost records has a substantial impact on the success of a building business.

Applying Cost Records

You must use careful judgment when applying recorded cost information to estimates for subsequent houses. The building conditions of two houses even of the same design are never the same because of different site and weather conditions, subcontractors, crews, and supervisors. Estimating from cost records is not an exact process.

The subcontract costs are usually bid individually for each house and, therefore, are not transferable from one house to the next unless you have worked out some scheme with your subcontractors by which you can bid their work on a unit-cost basis.

You should frequently update material costs for estimating purposes. Many vendors provide price books for estimators' reference that are valid for a stipulated period of time. These quotes eliminate the need to get prices on individual items for each estimate, provided that price lists are revised and that the prices are competitive. If you use material cost records, even in preliminary estimates, you need to consider a contingency cost for material price escalations.

Your records are the single source of cost information for overhead items. Logging all expenditures by jobsite cost-estimating categories develops a broad-based reference for each item. You need the record of time on site and of actual use of items to validate the quantities estimated for future projects. Applying time and rate of costs from one house to another requires judgment and experience to select appropriate cost factors.

Published Cost References

A number of cost reference books are available for estimating purposes. All of the sources qualify their cost information to make it more useful. The best references for residential builders are those specifically developed for residential takeoff and cost estimating.

Some references provide substantial support information for quantity takeoff, often with cost extensions. Tables similar to those in this book provide for a broad range of quantity conversions from measurements to pricing units. Most manuals specify qualified unit costs for labor, material, equipment, and subcontract items.

If you use published references, you must understand and carefully consider the factors involved and the qualifying of the costs listed. The labor unit costs furnished in the manuals specify crew size, wage rates, and local cost adjustment factors.

Although wage rates differ tremendously from one area to another, the expected productivity of a craftsperson or crew should be similar in residential construction. You can compare your labor productivity estimate to a national standard by adjusting the published unit cost factors by the wage rate you used.

Example—A brick masonry subcontractor might charge $1 per block and $300 for 1,000 bricks for labor. The national standard provides unit costs for brick masonry labor of $1.25 a block and $400 for 1,000 bricks. If the subcontractor's wage rate is 20 percent less than the published rates, the productivity of the subcontractor equals that of the published average.

Avoid using the published labor unit cost directly for your cost estimate. The estimating standards are best used as references for average crew productivity as indicated by the published unit cost. Many manual estimates are conservative, and you can use the costs listed for a number of similar work items to make adjustments before estimating costs.

The published material unit costs are averages and vary too much to use in cost estimating. However, the pricing units are standard, and the material breakdown can aid in developing a checklist or identifying all the requirements of a particular method of construction or a particular detail.

The published equipment and jobsite overhead cost information is also an average of costs. However, you can review and use the factors for crew size applied to equipment or those applied to jobsite overhead items to estimate building requirements and to help validate current practices.

Guidelines for Accuracy

Accuracy in cost estimating cannot be overemphasized. Building companies live, and often die, by their cost estimates. If your cost estimates are inaccurate because of overestimated costs or quantities, you will not be competitive. Inflated estimates make preconstruction selling and competitive bidding difficult. Such estimates also hinder useful cost control efforts.

An agreement to complete construction of a house for a specific lump-sum cost limits the funds available for the construction of that house. When estimates are less than the actual cost of construction, your profit must be used to offset the shortfall. You end up with less profit, no profit, or even a loss of out-of-pocket money.

You can be sure of your estimate if the quantity takeoff is complete and correct, but no foolproof method exists to prevent errors. However, standard practices will reduce the chance of error. Following the guidelines in Figure 5-12 will improve the accuracy of the quantity takeoff and cost estimate.

Figure 5-12. Techniques for Reducing Errors in Estimates

- Use stated dimensions for all calculations. Use scale drawings only when stated dimensions are unavailable.
- Measure everything as shown. Do not approximate, average, assume, or round off.
- Take off everything that is shown. Use a checklist to ensure that all items are found. Check off each item to ensure against double counting.
- Keep different items separate and group like items together. (The estimator must understand how to price the takeoff so that items with different material, labor, and/or equipment transportation costs are not inadvertently grouped together.)
- Show all units of measurement to avoid confusion and error.
- Round off only after completely calculating the quantity for each specific item. Round off to the next highest number, except for deducts; in which case, round off to the next lowest number.
- Label each entry on the takeoff by section number, location description, or item number for ease in reference and checking.
- Enter each item in the appropriate pricing unit. The cost extensions will be accurate only if the unit cost matches the units used in the quantity takeoff.
- To avoid error, be careful when writing the takeoff data on the takeoff forms. Check each entry for accuracy.
- Minimize the number of computations and cost extensions. Each calculation is an opportunity for error. Therefore, use conversion factors, tables, and other estimating references that reduce the number of computations required.
- Have one person check another's takeoffs, cost estimates, and arithmetic to resolve errors and omissions before the construction cost for the house is proposed.
- Save time by getting the answers to any questions about the plans in a single meeting or phone call after you review the drawings and specifications.

Avoiding Common Errors in Estimating

The worst type of error in construction cost estimating is leaving out an item of work. On a lump-sum bid and contract, you will be obligated to do all of the work, even if you have omitted the cost for part of the work in the estimate and in the contract to build.

Ensuring that all items have been included in the takeoff is perhaps the most difficult part of the estimate to do. The first step (before beginning the takeoff) is to read all of the notes and details on the plans. Underline, check, or circle any items that may be difficult or unusual in any way.

Although the process of checking off each item primarily prevents double counting, it also reduces the chance of omission. Systematically checking off each item facilitates reviewing plans at the end of the takeoff, especially if you pay particular attention to unchecked items that may have been missed.

One error that should never be made but often is, is leaving out the cost of work that has been taken off but is not included on the summary form when completing the estimate or bid. The best way of eliminating this error is through a comprehensive system to control the estimate summary process. Transfer a complete list of work or cost items to the cost estimate summary sheet before completing the estimate or bid. Check the cost estimate for completeness before using the estimate as a bid.

Transposed numbers are difficult to detect in measurements, quantities, or cost extensions. To eliminate transposed figures, check each and every entry at the time it is made. All computations required for the quantity takeoff and cost estimate should be figured on a calculator with a paper tape so you can easily compare every entry with the worksheets and recalculate as needed.

Chapter 6

The Cost Control System

The cost control system is based on measuring actual incurred costs and comparing them to the funds budgeted for the item of work, identifying any problems or mistakes, and reviewing and implementing alternative corrective actions. To realize profits on a consistent basis, you must use a systematic method of promptly and accurately collecting and using cost information. The initial objective of a cost management system is to control the cost so you do not exceed your budget estimate.

The available funds for the work items form the basis for planning the work and detecting problems. The cost estimate provides not only an overall budget, but it also lists the cost categories on an item-by-item basis. The complete estimate must include enough detail to support the cost control system.

The frequency of cost reports is different for different cost categories. A weekly payroll system can and should be used to monitor and control labor cost. Equipment cost depends on the amount of time it is used, a factor that can be monitored during site visits. Material and subcontract costs normally accumulate on a monthly basis. For this reason, you need separate and frequent loss and waste control on site.

The best method for controlling costs is to know the budget cost for each item and then to pay close attention to the expenditure of that budget. Control of construction costs, therefore, should be based upon your estimated cost on an item-by-item or individual work-category basis.

Example—If your estimate provides $2,640 to frame an addition to a house, and you plan to pay a crew $55 per hour, the budget provides up to 48 crewhours (6 crewdays) of work. If less than half of the task has been completed in 3 workdays, you must substantially increase crew productivity or incur a cost overrun.

The Estimate Breakdown

The estimate breakdown is used to determine the workhours, quantity of materials, and equipment time for each work item or category of work.

To transfer estimated costs to the actual job cost control system, the estimate categories must match the categories used in the job cost management system. The NAHB Chart of Accounts[1] cost code system provides an appropriate reference for developing compatible estimating and cost control categories.

The construction cost cycle is shown in Figure 6-1.

The breakdown budgets the work on an item-by-item, category-by-category basis. A cost control system must recognize the requirements of each step in the building process and respond appropriately.

Figure 6-1. Construction Cost Cycle

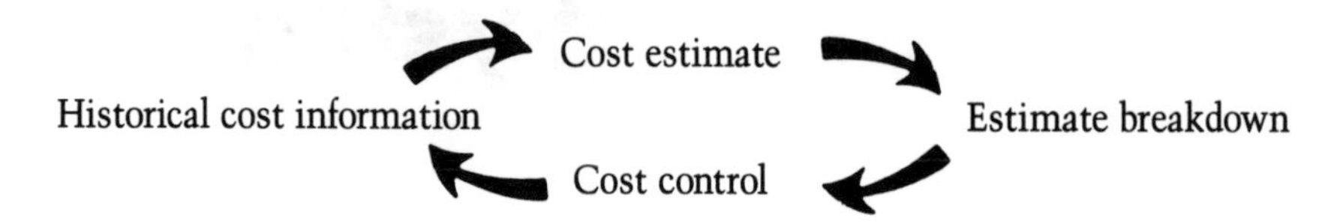

Labor Cost Control

Subcontracted labor will typically be estimated and budgeted as a lump sum—a specific amount of pay for a specific amount of work or unit price. Payment is made on a unit-completed basis or hourly fee. In managing lump-sum labor, your the greatest concern is monitoring the work completed versus any partial payment the subcontractor requests. For unit-price subcontractors, in addition to making sure that they are not overpaid for work in place if they get a draw, you also must be sure that you are paying them for the correct number of units.

Most builders' direct labor is limited to a few crafts or multicraft crews who work throughout the construction of a project. These builders manage their labor cost based on a single, broad-range cost objective. The total labor cost estimate breakdown is a time budget for the crew. This breakdown controls the labor cost by directing the work to be done within the budgeted time period. Often a builder groups labor cost estimates for individual work items with related items of work to establish the budget for a category of work.

Example—When all the work is to be performed by the same crew, you might add the line item labor cost estimates to complete the formwork for an on-grade slab with the estimates for the following as the actual slab labor budget:

- Excavating
- Installing the sand, gravel, fill, vapor barrier, and reinforcement
- Placing the concrete

In this case the total labor cost estimate divided by the crew cost per day creates the budget for the amount of time to complete that category of work.

When many different crews are used in constructing a single house, the actual production must be measured against the estimated production on an item-by-item basis to manage labor costs.

If workers' payroll time sheets are coded according to the NAHB codes in the NAHB Chart of Accounts,[2] the processing system accounts for the cost properly. These costs are segregated from material costs. Because labor costs are collected on a weekly basis, you need to process them weekly, so you can react to potential cost overruns in a timely manner.

Work items of less than a week's duration cannot be managed with this or other, similar labor cost management systems. Those labor cost estimate breakdowns result in too short a time period to accomplish the work, and you must evaluate the progress and productivity of the crew.

Example—If you develop a $600 cost estimate to install cabinets using a $25 an hour crew, you are budgeting 3 crewdays to complete the work. At the end of each day, you would have to evaluate the progress in order to make adjustments and corrections as required to meet the time budget and the labor cost estimate.

The standard measurements of labor are in workhours, crew hours, or hourly or daily crew costs. You can easily recognize your labor expenditure on a daily basis and, if you are aware of the estimated labor cost per work item, you can make adjustments to affect the total labor cost of the work item.

Equipment Cost Control

Equipment cost control should be separated from the control of other categories of cost. You can control equipment cost by controlling the time that equipment is used on the job site. Therefore, the control of that cost is similar to the control of indirect cost, except that the periods of equipment use relate to the time durations for particular items or categories of work, while indirect cost occurs throughout the building of the house.

Each piece of equipment has a time-based cost (See Chapter 2, "The Complete Estimate."). Therefore, you must carefully schedule the use of the equipment to fit the amount of time budgeted in the estimate before the equipment arrives on site. The most frequent equipment cost overrun results from equipment sitting idle on the jobsite before or after it is needed in the building sequence.

The starting date of the equipment charge determines the date that use of the equipment must be complete to prevent a cost overrun. The superintendent and foreman must be aware of this time frame and ensure that work with the equipment is finished within the allotted time.

Material Cost Control and Cost Summary

The material cost control system does not function identically to the labor cost control system. The invoices for materials list the number of units of each item purchased, a unit price for each item, and a total cost. When comparing the actual invoice to the estimated costs, not only the total cost, but also the number of units and the unit price must match. You should evaluate any discrepancies.

Most builders buy materials with a purchase order system. and issue purchase orders as soon as possible after signing a contract to build a house. The vendor quotes may have specific expiration dates. Thus, cost overruns can occur if purchase orders are delayed and unit prices increase. As long as the purchase order is issued within the time period specified in the quote, the cost overrun on materials that were estimated based on fixed quotes is limited to any difference between actual and estimated quantity.

When you actually place an order for material, you should verify any cost estimates based on prices at the time of bid. If a price has changed, you should record the cost overrun or underrun and compare it with the material contingency cost that was included in the cost estimate.

Most builders collect delivery tickets and compare the quantity of materials delivered to the quantity of materials listed on the invoice. Many builders also furnish a list of material quantities at the jobsite for controlling material orders or deliveries and compare the amount delivered to the quantity shown in the estimate. A substantial variance between the amount of materials paid for and the amount of materials estimated indicates either an inaccurate quantity takeoff or an excessive loss or waste of material at the job site. You need to identify and correct the problem to prevent cost overruns on future projects.

If an estimate for materials is based on a quote, the unit cost for those materials charged on an invoice must not exceed the amount listed in the purchase order. You have to compare the actual unit cost to the estimated or quoted unit cost to assure that the cost charged is appropriate. Whether inadvertent or intentional, you can prevent cost overruns caused by the vendor's billed unit cost varying from the quoted unit cost if you use caution and a systematic approach to the payment process.

Because the quantity of materials used and the quantity of materials estimated can vary substantially, a monthly cycle of invoice control prevents overcharges and facilitates payment within the discount period.

The materials must be ordered from a vendor charging no more than the unit cost used in the estimate, or a cost overrun will occur. A list of approved vendors should be available at the jobsite for materials ordered on site. You should not incur cost overruns from using more costly materials, regardless of vendor convenience.

Limiting the amount of material stored on site reduces the opportunity for theft, a frequent cause of material cost overrun. Needing more material to complete an item of work signals a variation from the estimated quantity. You can resolve the problem of inaccurate waste/loss factors in future estimates by accurately measuring the material quantities purchased and used for a specific quantity of work.

Control of Indirect Costs

The risk of cost overrun in the indirect and equipment cost categories is substantial. The cost management system must identify any problems as early as possible so that corrective steps can be taken.

Control of indirect costs based on the estimate requires you to measure each cost as you incur it and compare it to the estimated cost to determine if a cost overrun can be anticipated. The quantities of work and cost is expressed in units of time, so to control indirect costs is to monitor and control time.

Labor burden and sales tax computations are often included with the indirect cost estimate. However, you can control the cost of the labor burden through labor time management. Sales taxes will increase or decrease with the actual cost of materials purchased. Sales tax is a percentage of direct labor and material costs, so control involves controlling these costs. If a labor cost overrun occurs, an overrun in the labor burden category also occurs. If total material cost exceeds the total material cost estimate, an overrun in the sales tax category also occurs.

The cost accounting system how much of the estimated indirect cost should have occurred during a reporting period. You can compare this estimated amount to the actual indirect cost incurred. You can then anticipate cost overruns and take preventive actions.

Chapter 7

Computerized Estimating

Advantages of Computerized Estimating

A computerized estimating system can have a tremendous impact on the quality, speed, and accuracy of your work. These systems can relieve an estimator of time-consuming tasks by automatically carrying out extensions for waste factors, productivity rates, conversions, rounding, and price factors. Automating these tasks reduces the chance of common errors and increases the accuracy of the estimate.

Computerized estimating systems often have a database from which the estimator draws information for a particular estimate. A database eliminates much of the repetitive work associated with manual estimating. This change allows more time to review and qualify an estimate, provides the estimator opportunities to experiment with scenarios, and produces instant results.

This chapter offers an overview of computerized estimating systems for home builders, but it does not provide complete form or content. Individual company needs and goals must be discussed with a qualified consultant and/or software vendor before a builder or remodeler purchases software or hardware. A few of the benefits associated with computerized estimating systems are discussed immediately below.

Increased Productivity

In considering manual estimating methods and, particularly, the quantity takeoff, you must write down each item (for example, a 2″ × 4″ × 8′ stud) each time you run across it in each estimate. Using a computer could eliminate this repetitious work for each estimate. You would simply select the item from a list (database). This capability would save you considerable time and effort in the takeoff process.

Scope of Work

Many estimators refer to a checklist to identify the items in an estimate. These checklists are designed to cover the entire scope of an estimate and to provide some assurance that you have included everything required to complete the estimate. A computerized estimating system provides you with a similar checklist that enables you to first identify the scope of the estimate and later to calculate the appropriate quantities. This automation saves additional time during the takeoff process.

Accuracy

Every estimator can make mistakes, such as simple mathematical errors, transposed numbers, or miscounted decimal places. Mistakes can occur throughout the estimate, including counting and calculating quantities for the takeoff and extensions. A computerized estimating system can automate this process by using formulas to perform takeoff calculations and by extending the unit prices in the spreadsheet. Allowing the computer to accurately perform this work gives you more time to qualify the numbers and thus produce an even more accurate estimate. After the manual takeoff was completed, a computerized estimating system was used to estimate the house plans in Chapter 3, "The Quantity Takeoff Process," (Figures 3-4 to 3-11). This system identified mathematical errors in the manual estimate, which have been left unchanged. (Refer to the "Interior Paint" item under the "Paint and Wallcover" section of Figure 4-19 for more information.

Standardization

Estimators often compare one estimate to another to check unit prices, dollar amounts, and other items. More often than not, the format changes from one estimate to the next. This inconsistency makes comparing estimates difficult because estimates are not organized the same way. (You are not comparing "apples to apples.") A computerized estimating system offers standardized estimates and provides a basis to compare previous work to new work.

Other benefits of standardization include the following consistent elements:

- Provides standard estimating methods for your company to follow.
- Makes estimates that are easy for others to review.
- Improves the organization of your work.

Contingency Scenarios

To arrive at the best total estimated amount, an estimator needs the ability to manipulate the numbers (for such items as unit prices, dollar amounts, add-ons). This manipulation is a nightmare without a computer and, thus, is rarely attempted with manual estimating methods. A computerized estimating system allows what-if possibilities, including changing unit prices, dollar amounts, productivity rates, percentages of waste, workhours, add-ons, or markups. Most importantly a computerized system enables you to immediately see the results of these changes, without affecting the estimate, until you are satisfied with the results. An efficient computerized estimating system will have the advantages discussed below.

Approaches

Builders and remodelers can develop many different ways to estimate a project, and they have just as many different ways to develop a computerized estimating system. Computerized estimating generally is based upon two different approaches: spreadsheets or databases. Each has its advantages and disadvantages. A good estimating system will incorporate the best of both approaches and add a few features of its own.

Generic Spreadsheets

Generic spreadsheets are fast, free-form, open systems that require considerable initial setup. The estimator must define and format each cell and subtotal and ensure that the information is entered correctly. All data must be reentered for each estimate because no database exists. Sometimes in these open, free-form systems some information (i.e., a unit price or a dollar amount) is entered in the wrong cell and not added to the subtotal. Builders commonly use these spreadsheets for summarizing an estimate, after the manual takeoff has been completed.

Spreadsheets provide the following advantages:

- Ease of use
- Fast, immediate results
- Familiar to most estimators
- Inexpensive
- Free-form

Spreadsheets also include these disadvantages:

- Considerable initial setup
- Potential inconsistency
- Not specifically estimating-oriented
- No predefined estimating reports
- Limited takeoff capabilities

Dedicated Estimating Systems

Dedicated estimating systems usually depend upon a database of information that stores such information as items, unit prices, production rates, and percentages of waste. These systems are specifically designed for construction estimating and may or may not require initial setup. They typically are not as fast as spreadsheets, but they offer database capabilities, automatic calculation features, and standardized estimate formats. Dedicated estimating systems incorporate a consistent, methodical data-entry process and often require the estimator to print a report to see the results.

Dedicated estimating systems provide these advantages:

- Estimating specific
- Powerful takeoff capabilities
- Database capabilities
- Consistency

Dedicated estimating systems include the following disadvantages:

- Slow, latent results
- Unfamiliar to estimator
- May be incompatible with company estimating procedures
- Difficult to learn
- More costly

Spreadsheet-Oriented Estimating Systems

These systems incorporate the familiar, open, free-form approach to the generic spreadsheet that enables you to change prices, productivity rates, and dollar amounts on the screen and immediately see the results. Spreadsheet-oriented estimating systems also incorporate a database and provide powerful takeoff capabilities, checklist features, and consistent results. These systems are easy to use and usually cost more than generic spreadsheets but less than dedicated estimating systems.

Discussion of these features and capabilities of spreadsheet-oriented estimating systems appears in the next sections. Many of the features and capabilities of these systems may apply to generic spreadsheets or dedicated estimating systems, but not to both.

All future references to estimating systems specifically identify spreadsheet-oriented estimating systems unless stated otherwise.

Database Capabilities

A database is similar to a price book that contains the estimator's knowledge, including conversion and waste factors, unit prices, and typical items used in an estimate. The database is the foundation of a computerized estimating system.

Standards

Each company develops standardized data that remains consistent for each estimate. This consistency provides a basis for comparing estimates as well as consistent estimate formats.

Standard cost codes or phases (such as 6000 Framing or 9000 Finishes) provide a well-organized coding structure that typically does not change for each estimate, and this structure should integrate with the company cost control system. Cost codes or phase numbers can be based upon industry standards, such as, those of the Construction Specifications Institute (CSI), the National Association of Home Builders (NAHB), or an internal company system.

Spreadsheet-oriented systems can maintain a builder's or remodeler's subcontractor and vendor lists, including names, addresses, contacts, and phone numbers. These systems can organize (a) subcontractors by their type of work for bidding purposes and (b) vendors by the materials they supply to handle purchasing or material price quotas. These systems also can produce mailing labels.

Items

The items entered maintain the detail of the database such as 2″ × 4″ × 8′ studs, footings, concrete, or roofing felt. The items store information such as unit prices, productivity rates, and waste conversion factors. Conversion factors will convert takeoff quantities to order quantities.

Example—These systems can use a conversion factor to calculate rolls (order quantity) of vapor barrier if the estimator enters the square feet (takeoff quantity) of slab area.

Formulas

Formulas assist in calculating quantities of items for takeoff purposes. They reduce the risk of errors, such as pressing the wrong key on the calculator because the arithmetic functions are performed automatically.

Example—A formula such as "Volume—CY" will ask for the length, width, and depth in feet. It will multiply the three values and divide by 27 to covert cubic feet to cubic yards. As you create formulas, you store them in the database, and they are accessible during takeoff.

Work Packages

Work packages (assemblies) are logical groups of items that make up a particular structure, such as a wall or slab. They cross the boundaries of the database numbering structure.

Example—A footing work package may contain excavation, concrete, and steel items. These items are located in different divisions or categories (based upon CSI's Masterformat®). A work package can group these items together, rather than select them separately, and apply one common set of dimensions (such as, length, width, excavation depth, and concrete depth) and calculate the correct quantities for each item.

Add-Ons

Add-ons or markups are indirect costs, such as sales tax, general overhead, and profit. Add-ons are maintained in a master list from which the estimator selects those needed for each estimate. These add-ons maintain default or minimum rates (general overhead 5%) or amounts (fee $3,000). They can also mark up specific cost categories, such as labor burden for the labor cost. The default rates can be adjusted for each estimate.

Add-ons can be allocated (or buried) proportionately across the estimate. This capability is essential for negotiating purposes because it keeps you from disclosing your markups and fees to the party sitting across the table.

Example—You could allocate the general overhead and profit (across the entire estimate), Worker's Compensation (across the labor dollars), and retain the sales tax in the estimate totals page.

Crews

Labor and equipment cost classes (such as carpenters, laborers, and spray equipment) are setup in the database with their respective hourly rates and benefits (such as payroll taxes and Worker's Compensation). These classes can be assigned to a specific crew such as the framing or the finishing crew. The classes and/or crews can then be assigned to any item in the database. Similarly a crew can be selected for an item while in the estimate. The estimating system will automatically calculate the average workhour rate or combined crewhour rate.

Maintenance

Database maintenance involves the periodic updating of item prices. This update can be done on an individual item basis, such as listing the material prices and updating the appropriate items. You can also update the item prices globally by a specified percentage rate. Maintenance of item prices may be based upon a price update code.

Example—A home builder or remodeler may purchase lumber materials from a specific supplier, Woody's Woodworks. The estimator can assign the price update code "WOODY" to all the desired lumber items. By doing so, the estimator can either call up just the "WOODY" lumber items and update prices manually or change all the "WOODY" items globally by a specified percentage rate.

Takeoff Capabilities

A computerized estimating system enables you to build an estimate directly on the spreadsheet screen. To add items to the spreadsheet, you open a takeoff window, similar to a quantity takeoff. You also can select items in the database from a help list. Two basic methods of takeoff are possible: checklist takeoff and work package takeoff.

Checklist Takeoff

Checklist takeoff is the process of individually selecting items that need to be included in the estimate. You have these options:

- Selecting all the items in a cost division (such as lumber items) on a page or in an entire estimate and calculating the appropriate quantities.
- Selecting items by entering the appropriate quantities for the desired items. If the quantity is not known at the time of takeoff, the item is simply selected, and the quantity is calculated later. Item quantities can be calculated by plugging in a value, or by using an on-screen calculator or formula.

Work-Package Takeoff

Work-package takeoff is the process of selecting a predefined structure or object, such as a stud wall, a footing, or floor framing. The estimating system automatically selects all the items that may be required for that object or structure.

Example—A stud-wall work package would include items such as plates, studs, blocking, nails, sheetrock, and insulation. Once you select the items, you enter the dimensions such as length, wall height, and stud spacing in feet. The estimating system calculates the correct quantities for the appropriate items, based upon the dimensions entered. The work package also summarizes the total dollars for all the categories (such as labor, materials, and subcontracts) for all the items as well as the total unit cost for the stud wall. This capacity enables you to review the calculated values and spot any errors immediately. This process also improves the overall accuracy of the estimate. Work packages can also be set up as options.

Example—A home builder with preset floor plans of components (such as garages, decks, and sunrooms) may define all the options as work packages and assign *actual* quantities for each item in that option. If a garage consists of 80 studs, 200 linear feet of plate material, and other materials, these items and quantities can be defined in the garage work package and automatically generated when the option is selected.

Estimating Capabilities

Once items are selected in the takeoff process, they are generated to the spreadsheet and extensions are automatically calculated. Once they are in the spreadsheet, you have the ability to review the calculated quantities and amounts and manipulate the information as the estimate conditions require. All changes affect the individual estimate, but they leave the database intact.

Prices

As an estimator, you should always review the unit price of each item in the spreadsheet. You can change unit prices immediately and see the calculated dollar amounts. Similarly, you can change dollar amounts, and the system will recalculate the unit price. You may also wish to add dollars to an item.

Example—For installation purposes, an item may require a special piece of equipment (with a daily rental rate of $100). Typically, only labor and material prices are associated with this item. You simply add $100 to the equipment column for this estimate only.

Detail

Each item has additional details that may not appear on the spreadsheet screen. These details include information such as the crew size, waste factor, and productivity rate. The system provides a detail window that allows you to make any necessary changes or explore what-if scenarios without affecting the spreadsheet. The detail window allows you to see the results before deciding to make the change permanent.

You can also add memos about each item, such as any assumptions, reasons for changing a value, or just notes to yourself or others. Many items in the reports have memos that appear directly below the descriptions.

Organization

The spreadsheet is typically organized by cost code or phase number sequence and then by item code. The system provides subtotal information for each phase, including dollars and total labor and equipment hours. This organization may not be appropriate for all estimators at all times.

Alternatively, if you wish to review the items based upon their location, such as by floor number, the spreadsheet sequence can be altered to display the items by floor. Or, in this case, a work-item field was created to organize the computerized estimate to reflect the manual estimate.

Estimate Totals

At any time, you have instant access to current estimate totals. This access includes all direct category costs (such as labor, materials, subcontractors, equipment, and other) as well as any add-ons you have selected, such as sales tax and profit. The estimating system can also calculate a unit cost for the entire estimate. You can adjust add-on rates or dollar amounts for each particular estimate. Once an estimate is complete, you may want to adjust the final dollar amount or unit cost (cost per square foot).

Example—The system calculates a unit cost of $52.80 per square foot for a 1,702-square foot house. You believe the market rate is $55 per square foot. You can enter the desired $55 per square foot rate, and the system immediately displays a difference of $3,610.

You can decide where the additional dollars should be allocated (or buried). You could proportionately spread the additional dollars in the labor category, put the entire amount in the electrical subcontractor's bid, or a combination of the two. Most importantly, you also can return the estimate to the initial value of $52.80 square feet. Only with the aid of a computer and a spreadsheet-oriented estimating system are what-if scenarios possible.

Reporting

Reporting capabilities are one of the distinct advantages in purchasing an estimating system. Unlike generic spreadsheets, that have limited reporting capabilities ("What you see is what you get."), estimating systems offer such powerful features as multiple types of reports and sorting orders.

A sample report has been included in the sample estimate (Figure 7-1) for your review. This report is sorted by work item. You will notice that this report includes columns for labor, material, subcontractors, and other. The "Job Overhead" items were assigned to the "Other" column because they are not true subcontractor items (work performed by others). By using the "Other" category, the computerized estimate displays labor and materials (direct costs incurred by the home builder), subcontracted costs (work performed by others), and other costs (intangibles that are the home builder's responsibility). This capability produces a more accurate cost breakdown for the estimate—another advantage of a computerized estimating system. Other types of reports are discussed below.

Audit Trail—A printout of all the calculations generated in the estimate including use of the calculator and/or formula feature.

Standard Estimate Report—This report can be sorted by phase, location, work item, and other elements. It provides subtotals for these elements as well as grand totals for the estimate. Multiple levels of detail are available, depending on how much (or how little) information you want to print.

Bill of Materials—This report is sorted by material class, and it groups items to facilitate the purchasing process. A bill of materials, sorted by location, enables you to purchase materials by floor. A bill of materials quote sheet can be printed and sent to suppliers to solicit unit prices based upon specific quantities. (A complete, integrated purchasing process is described in the next section under "Purchase Orders and Subcontracts.")

Custom Reports—A custom report writer can generate a user-defined report format that may not be available in the estimating system.

Discrepancies between the manual estimate versus the computerized estimate result from differences in rounding methods.

Figure 7-1. Sample Estimate

Description	Wood Chase S/D	House Area in Sqft	1,702	Report format	Work Item order
	Lot 2	House Cost	$62,556		Detail report
	Block C	Calculation Method	$62,556/1,702		Print crews
		House Cost/Sqft	$36.75		Print memos
		Total Cost	$89,995		
		Calculation Method	$89,995/1,702		
		Total Cost/Sqft	$52.88		

Timberline Software, Inc. — Estimating Std Report by WBS — Wood Chase S/D — 9-09-91 12:14 pm — Page 1

ITEM DESCRIPTION	TAKEOFF QTY		<---LABOR--- UNIT PRICE	---LABOR---> AMOUNT	<---MATRL--- UNIT PRICE	---MATRL---> AMOUNT	<---SUB--- AMOUNT	---SUB---> NAME	OTHER AMOUNT	TOTAL AMOUNT
1. JOB OVERHEAD										
---- Superintendent	1.00	ls	3,333.00 /ls	3,333	-	-	-		-	3,333
$40,000/year salary / 12 houses per year										
---- Temp Electricity Hkp	1.00	ls	-	-	-	-	-		15	15
---- Temporary Electricty	1.00	ls	-	-	-	-	-		80	80
---- Water Tap Fee	1.00	ls	-	-	-	-	-		255	255
---- Sewer Tap Fee	1.00	ls	-	-	-	-	-		600	600
---- Appraisal	1.00	ls	-	-	-	-	-		200	200
---- Sales Comission	1.00	ls	-	-	-	-	-		5,400	5,400
$90,000 * 6%										
---- Cnstrctn Ln Clsn Cst	1.00	ls	-	-	-	-	-		220	220
Loan Origination Fee, Title Exam, Deed Prep, Record Deed, Closing Fee										
---- Sales Closing Cost	1.00	ls	-	-	-	-	-		1,350	1,350
Loan Origination Fee, Document Prep, Photos Closing Fee, 2nd Appraisal										
---- Lot Closing Costs	1.00	ls	-	-	-	-	-		200	200
Record Deed, Title Exam, Closing Fee, Taxes										
---- Plans	1.00	ls	-	-	-	-	-		300	300
---- Builder's Rsk Insrnc	1.00	ls	-	-	-	-	-		100	100
---- Liability Insurance	1.00	ls	-	-	-	-	-		35	35
---- Completed Operations	1.00	ls	-	-	-	-	-		50	50
---- Interest	1.00	ls	-	-	-	-	-		1,200	1,200
$20,000 * 1% 1st month										
$40,000 * 1% 2nd month										
$60,000 * 1% 3rd month										
---- Permits	1.00	ls	-	-	-	-	-		195	195
$15 Plan Review + $3/1000 Construction Costs										
---- Inspections	3.00	ea	-	-	-	-	-		90	90
	1. JOB OVERHEAD			3,333*					10,290*	13,623
2. SITEWORK										
---- Mobilization	1.00	ls	-	-	-	-	300		-	300
---- Site Grading	8.00	hr	-	-	-	-	480		-	480
	2. SITEWORK						780*			780
3. FOOTINGS & SLABS										
---- Footing Concrete	15.00	cy	-	-	55.00 /cy	825	-		-	825
Exterior Footings										
---- Footing Concrete	1.00	cy	-	-	55.00 /cy	55	-		-	55
Interior Footings										
---- 2x4x8 Batter Boards	16.00	ea	-	-	1.65 /ea	26	-		-	26
---- 2x8x14 Form Boards	13.00	ea	-	-	6.34 /ea	82	-		-	82
---- 2x6x12 Brickledge	3.00	ea	-	-	4.16 /ea	12	-		-	12
---- 2x4x8 Wood Stakes	25.00	ea	-	-	1.65 /ea	41	-		-	41
---- 2x4x8 Braces	12.00	ea	-	-	1.65 /ea	20	-		-	20

Figure 7-1. Sample Estimate (continued)

ITEM DESCRIPTION	TAKEOFF QTY		<-----LABOR----- UNIT PRICE	AMOUNT	<-----MATRL----- UNIT PRICE	AMOUNT	<----SUB----> AMOUNT	NAME	OTHER AMOUNT	TOTAL AMOUNT
---- 2x4x12 Garage Forms	2.00	ea	-	-	2.64 /ea	5	-		-	5
---- Ftng Rebr #3, Grd 40	23.00	ea	-	-	2.65 /ea	61	-		-	61
---- 1/2" x12" Anchr Blts	30.00	ea	-	-	.60 /ea	18	-		-	18
---- #5 Smt Dow 2'-0" Lng	14.00	ea	-	-	.40 /ea	6	-		-	6
---- S.O.G. Concrete	21.00	cy	-	-	55.00 /cy	1,155	-		-	1,155
4" Thick Slab										
---- 6x6- W1.4 x W1.4 W M	1,702.00	sf	-	-	42.00 /rl	126	-		-	126
---- 6 mil Vapor Barrier	1,800.00	sf	-	-	.023/sf	41	-		-	41
---- S.O.G. Gravel	21.00	cy	-	-	6.242/tn	206				206
4" Thick gravel										
---- Form & Pour Concrete	1,702.00	sf	-	-	-	-	1,191		-	1,191
	3. FOOTINGS & SLABS					2,680*	1,191*			3,872
4. MASONRY										
---- Brick Veneer	2.75	m	-	-	200.00 /m	550	575		-	1,125
Walls: 27*2.5*7 bricks/sqft=473										
Fireplace: 19*16*7 bricks/sqft=2128										
---- Mortar Mix	20.00	sk	-	-	4.50 /sk	90	-		-	90
@ 140 bricks per sack										
---- Sand	3.00	cy	-	-	20.00 /cy	60	-		-	60
@ 1 cuyd per 1000 brick										
---- Wall Ties	1.00	bx	-	-	12.00 /bx	12	-		-	12
---- Firplc w/Chmny (prf)	1.00	ea	-	-	689.00 /ea	689	-		-	689
			4. MASONRY			1,401*	575*			1,976
5. FRAMING										
---- 2x8 Pressr Trtd Sill	150.00	lf	-	-	.54 /lf	81	-		-	81
---- 2x4 Pressr Trtd Sill	250.00	lf	-	-	.27 /lf	68	-		-	68
---- 2x4 Precut Studs	560.00	ea	-	-	1.55 /ea	868	-		-	868
---- 2x4 Plates & Blockng	1,200.00	lf	-	-	.21 /lf	252	-		-	252
---- 2x12 Headers	180.00	lf	-	-	1.03 /lf	185	-		-	185
---- 20'- 0" Common Truss	10.00	ea	-	-	40.00 /ea	400	-		-	400
---- 20'- 4" Scissor Trss	5.00	ea	-	-	40.00 /ea	200	-		-	200
---- 14'-10" Scissor Trss	3.00	ea	-	-	28.00 /ea	84	-		-	84
---- 10'- 4" Scissor Trss	2.00	ea	-	-	21.00 /ea	42	-		-	42
---- 19'- 9" Stub Truss	18.00	ea	-	-	40.00 /ea	720	-		-	720
---- 13'- 4" Stub Truss	6.00	ea	-	-	27.00 /ea	162	-		-	162
---- 19'- 9" Gable Truss	1.00	ea	-	-	40.00 /ea	40	-		-	40
---- 14'-10" Gable Truss	1.00	ea	-	-	30.00 /ea	30	-		-	30
---- 13'- 4" Gable Truss	1.00	ea	-	-	26.00 /ea	26	-		-	26
---- 9'- 8" Gable Truss	1.00	ea	-	-	20.00 /ea	20	-		-	20
---- 2x6x16 Rafters	17.00	ea	-	-	5.25 /ea	89	-		-	89
---- 2x6 Rake Framing	250.00	lf	-	-	.30 /lf	75	-		-	75
---- 2x4 Lookouts/Ledgers	120.00	lf	-	-	.23 /lf	28	-		-	28
---- 2x4 Bracing	300.00	lf	-	-	.23 /lf	69	-		-	69
---- 2x12x18 Wood Beams	2.00	ea	-	-	16.15 /ea	32	-		-	32
---- 1/2"x4x8 Wall Shthng	31.00	sh	-	-	2.45 /sh	76	-		-	76
---- 1/2"x4x8 CDX Wal Sht	16.00	sh	-	-	6.40 /sh	102	-		-	102
Plywood Corners										
---- 1/2"x4x8 CDX Rf Shth	66.00	sh	-	-	8.55 /sh	564	-		-	564
---- Plywood Clips	1.00	bx	-	-	13.00 /bx	13	-		-	13
---- Roofing Felt	6.00	rl	-	-	9.00 /rl	54	-		-	54
---- 1x6 Rdwd Sidng (Bvl)	4,000.00	lf	-	-	.55 /lf	2,200	-		-	2,200
---- 1x6 Redwd Crner Brds	180.00	lf	-	-	.60 /lf	108	-		-	108
---- 3-0x6-8 Wood Door	1.00	ea	-	-	125.00 /ea	125	-		-	125
---- 2-8x6-8 Wood Door	1.00	ea	-	-	125.00 /ea	125	-		-	125
---- 6-0x6-8 Sldng Gls Dr	1.00	ea	-	-	315.00 /ea	315	-		-	315
---- 16-0x7-0 Garage Door	1.00	ea	-	-	785.00 /ea	785	-		-	785
---- 28x32/32 Sng-Hng Wnd	2.00	ea	-	-	132.00 /ea	264	-		-	264
---- 28x28/28 Sng-Hng Wnd	3.00	ea	-	-	124.67 /ea	374	-		-	374
---- 28x16/16 Sng-Hng Wnd	1.00	ea	-	-	133.00 /ea	133	-		-	133
---- 28x32/32 Dbl-Hng Wnd	1.00	ea	-	-	157.00 /ea	157	-		-	157

Figure 7-1. Sample Estimate (continued)

ITEM DESCRIPTION	TAKEOFF QTY	<---------LABOR---------> UNIT PRICE	AMOUNT	<---------MATRL--------- UNIT PRICE	AMOUNT	><------SUB-----> AMOUNT	NAME	OTHER AMOUNT	TOTAL AMOUNT
---- 28x28/28 Dbl-Hng Wnd	2.00 ea	-	-	138.00 /ea	276	-		-	276
---- Exterior Door Hardwr Locksets	2.00 ea	-	-	26.00 /ea	52	-		-	52
---- Window & Door Trim 1x6 Redwood	200.00 lf	-	-	.60 /lf	120	-		-	120
---- Louvered Vents	4.00 ea	-	-	51.00 /ea	204	-		-	204
---- 6x6x8' Redwood Posts Entry Post	1.00 ea	-	-	28.80 /ea	29	-		-	29
---- Nails Framing	1.00 ls	-	-	200.00 /ls	200	-		-	200
---- 2x8 Sub-Facia	100.00 lf	-	-	.44 /lf	44	-		-	44
---- 1x8 Redwood Fac Bard	250.00 lf	-	-	.80 /lf	200	-		-	200
---- 1/2"x4x8 AC Plyw Sof	5.00 sh	-	-	10.50 /sh	53	-		-	53
---- 1" Cnt. Soff Scr Vnt	90.00 lf	-	-	.50 /lf	45	-		-	45
---- 1x6 Redwood Frze Brd	90.00 lf	-	-	.60 /lf	54	-		-	54
---- 4x4x8' Prssr Trt Pst	4.00 ea	-	-	4.16 /ea	17	-		-	17
---- 2x4x8' Prssr Trt Bms	6.00 ea	-	-	2.08 /ea	12	-		-	12
---- 1x6x6' Prssr Trt Fnc	44.00 ea	-	-	2.40 /ea	106	-		-	106
---- Framing Labor (Hous)	1,702.00 sf	1.80 /sf	3,064	-	-	-		-	3,064
		5. FRAMING	3,064*		10,278*				13,341
6. ROOFING									
---- Metal Drip Edge	222.00 lf	-	-	.15 /lf	33	-		-	33
---- Valley Flashing	50.00 lf	-	-	.64 /lf	32	-		-	32
---- Asphalt Shingles	21.00 sq	-	-	24.00 /sq	504	231		-	735
---- Roofing Nails	1.00 bx	-	-	36.00 /bx	36	-		-	36
		6. ROOFING			605*	231*			836
7. PLUMBING									
---- Plumbing Sub (House)	1.00 ls	-	-	-	-	2,600		-	2,600
---- Tempered Glass Doors	1.00 ea	-	-	220.00 /ea	220	100		-	320
---- Watr & Sewr Connctns	1.00 ls	-	-	-	-	360		-	360
		7. PLUMBING			220*	3,060*			3,280
8. ELECTRICAL									
---- Electrical Sub (Hse)	1.00 ls	-	-	-	-	1,600		-	1,600
---- Lighting Fxtr Allwnc	1.00 ls	-	-	1,000.00 /ls	1,000	-		-	1,000
		8. ELECTRICAL			1,000*	1,600*			2,600
9. HVAC									
---- HVAC Sub (House)	2.50 tn	-	-	-	-	2,015		-	2,015
		9. HVAC				2,015*			2,015
10. INSULATION									
---- 12" R-38 Batt Ins Attic	1,000.00 sf	-	-	-	-	660		-	660
---- 3-5/8" R-13 Batt Ins Walls	1,920.00 sf	-	-	-	-	614		-	614
---- 9" R-30 Batt Ins Vaulted Ceiling	320.00 sf	-	-	-	-	166		-	166
---- R-7 Rigid Insltn Brd	320.00 sf	-	-	-	-	58		-	58
		10. INSULATION				1,498*			1,498
11. DRYWALL									
---- 5/8" Firecode Shtrck Garage	930.00 sf	-	-	8.65 /bd	174	-		-	174
---- 5/8" Regular Shtrck House	4,970.00 sf	-	-	7.78 /bd	806	-		-	806
---- Drywall Sub (House) Hang & Finish	124.00 br	-	-	-	-	1,309		-	1,309
		11. DRYWALL			980*	1,309*			2,289

Figure 7-1. Sample Estimate (continued)

ITEM DESCRIPTION	TAKEOFF QTY		<---LABOR---> UNIT PRICE	AMOUNT	<---MATRL---> UNIT PRICE	AMOUNT	<---SUB---> AMOUNT	NAME	OTHER AMOUNT	TOTAL AMOUNT
12. INTERIOR TRIM										
---- Wood Base	440.00	lf	-	-	.32 /lf	141	-		-	141
---- Wood Casing	530.00	lf	-	-	.27 /lf	143	-		-	143
---- Flush Wood Door	7.00	ea	-	-	65.00 /ea	455	-		-	455
---- Bifold Wood Door	7.00	ea	-	-	65.00 /ea	455	-		-	455
---- Interior Door Hardwr	1.00	ls	-	-	167.00 /ls	167	-		-	167
5 - Privacy Locks										
7 - Bifold Pulls										
9 - Bumpers										
---- 1x12 Whit Pin Shlvng	40.00	lf	-	-	.76 /lf	30	-		-	30
---- Wood Closet Rods	26.00	lf	-	-	.37 /lf	10	-		-	10
---- Oak Mantle Trim	1.00	ls	-	-	60.00 /ls	60	-		-	60
1x2x8 Oak, 1x6x12 Oak, 1x4x18 Oak										
3" Oak Crown Moulding - 14 lnft										
Oak Bead Moulding - 24 lnft										
---- Nails	1.00	ls	-	-	150.00 /ls	150	-		-	150
Interior Trim										
---- Inter Trm Labr (Hs)	1,274.00	sf	1.10 /sf	1,401	-	-	-		-	1,401
---- Bath Accssrs (Cmplt)	2.00	ea	-	-	62.50 /ea	125	-		-	125
Paper Holders, Soap Dishes,										
Toothbrush Holders, & Towel Bars										
---- Shower Rods	2.00	ea	-	-	11.00 /ea	22	-		-	22
---- Bath Mirror	32.00	sf	-	-	3.00 /sf	96	-		-	96
2 mirrors @ 4'x4'										
---- Dryer Vent w/Hose	1.00	ea	-	-	20.00 /ea	20	-		-	20
Includes 20' hose										
12. INTERIOR TRIM				1,401*		1,874*				3,275
13. PAINT/WALLCOVER										
---- Exterior Paint	4,720.00	sf	-	-	-	-	944		-	944
---- Interior Paint	1,273.00	sf	-	-	-	-	637		-	637
Manual estimate is incorrect; numbers are transposed										
Error in extensions										
---- Wall Paper	18.00	rl	-	-	-	-	360		-	360
13. PAINT/WALLCOVER							1,941*			1,941
14. FLOOR COVERING										
---- Interior Carpeting	100.00	sy	-	-	-	-	1,500		-	1,500
---- Vinyl Flooring	350.00	sf	-	-	-	-	700		-	700
14. FLOOR COVERING							2,200*			2,200
15. CABINETS										
---- Kitchen & Bth Cabnts	1.00	ls	65.00 /ls	65	1,508.00 /ls	1,508	-		-	1,573
15. CABINETS				65*		1,508*				1,573
16. APPLIANCES										
---- Range	1.00	ea	-	-	625.00 /ea	625	-		-	625
---- Refrigerator/Freezer	1.00	ea	-	-	850.00 /ea	850	-		-	850
---- Dishwasher	1.00	ea	-	-	290.00 /ea	290	-		-	290
16. APPLIANCES						1,765*				1,765
17. EXTERIOR										
---- 4" Concrete Sidewalk	80.00	sf	-	-	1.50 /sf	120	-		-	120
5 panels @ 4'x4'										
---- 4" Concrete Patio	130.00	sf	-	-	-	-	260		-	260
---- 4" Concrete Driveway	12.00	cy	-	-	55.00 /cy	660	288		-	948
---- Final Grading	8.00	hr	-	-	-	-	360		-	360
---- Trees & Shrubs	18.00	ea	-	-	-	-	270		-	270
---- Seed & Sod Grass	11,170.00	sf	-	-	-	-	819		-	819
17. EXTERIOR						780*	1,997*			2,777

Figure 7-1. Sample Estimate (continued)

ITEM DESCRIPTION	TAKEOFF QTY		LABOR UNIT PRICE	LABOR AMOUNT	MATRL UNIT PRICE	MATRL AMOUNT	SUB AMOUNT	SUB NAME	OTHER AMOUNT	TOTAL AMOUNT
18. CLEAN										
---- Inside Cleanup	1,670.00	sf	-	-	-	-	251		-	251
---- Outside Cleanup	3.00	tr	-	-	-	-	600		-	600
		18. CLEAN					851*			851

```
                  ESTIMATE TOTALS

                     7,863  Labor
                    23,091  Material
                    19,248  Subcontractor
                    10,290  Other
-------------
       60,492
                     1,616  Sales Tax                  C    7.00000%
                       448  Taxes & Ins On Labor       C    5.70000%
-------------
       62,556
                    15,700  Land Costs                 L
-------------
       78,256
                     3,913  General Overhead           T    5.00000%
                     7,826  Profit                     T   10.00000%
-------------
       89,995  TOTAL ESTIMATE  52.88/Sqft
```

Integration

A computerized estimating system provides a means of integrating estimating data with design, scheduling, cost control, and purchasing systems. An electronic exchange of data can enhance productivity, reduce the repetition of work, and assist in the exchange of relevant information.

Integration of Computer-Aided Design (CAD)

Computer-aided design software is to the designer what computerized estimating software is to the estimator. Designers use this tool to create electronic drawings with the assistance of the computer. Integration of CAD involves performing the takeoff process of an estimate right on the computer screen.

In this simplified process the estimator or designer selects objects in the CAD drawing as either items or work packages from a list similar to the estimating database list. The estimating database available in the CAD drawing software enables the estimator to work with a similar estimating system.

Example—The outline of a house could be selected as the slab work package. The CAD system automatically calculates the area, slab thickness, and perimeter.

Once complete, this information is sent (exported) to the estimating system, which automatically calculates the unit prices, applies productivity rates, and other data. While in the estimating system, the estimator has all the same capabilities for manipulating the data.

If the drawing changes, the information is reexported and the estimate is automatically updated to match the design changes.

Scheduling Integration

Computerized scheduling incorporates three basic steps:

- Identify the activities.
- Apply a duration to those activities.
- Define the logical relationships between those activities.

You can also apply additional information such as resources (crews or quantities) and budgets to each activity.

Integration of an estimating system with a scheduling system enables the scheduler to review the estimate information and identify which items will make up an activity (Step 1). Once an activity is defined, if the selected items contain either labor or equipment hours, the activity's duration is automatically calculated (Step 2). If not, the scheduler can assign the desired duration. Resources and budgets are also automatically set up for each activity.

The information is then sent to the scheduling system where the logical relationships are defined (Step 3). All additional work is performed in the scheduling system.

Cost Control Integration

Integrating a cost control system requires that you coordinate estimating phase numbers with cost control codes. The estimating data is exported to the cost control system to set up initial budgets for the job. You can also send quantities (workhours or square feet) if you plan to track quantities as well as dollars. (See also Chapter 6, "The Cost Control System.")

Purchase Orders and Subcontracts

Purchase orders and subcontracts can be generated from the estimating data through an intermediate step or buy-out process. Once the estimate is complete, the data is sent to the buy-out system. The buy-out process involves organizing the data and generating quote forms to solicit quotations from vendors. As quotes come in you can compare prices with the estimate and other vendors' quotes. At any time the system provides information such as the following:

- Which items have been or still need to be quoted
- Which vendor has the lowest price
- Which items have been or still need to be purchased
- How much the project is over or under the budget

Once you have finalized prices and selected vendors, the system can automatically produce purchase orders or subcontracts. You can check invoices against the purchase order or the subcontract.

To integrate software you must first develop the two systems you want to integrate, and you have to know the individual systems well before you can make them communicate.

Selecting Estimating Software

Selecting the estimating system that best meets your company needs usually requires an extensive, time consuming, evaluation process. You should review as many systems as possible until you find one with the capabilities you desire. Figure 7-2 lists a few key points to consider when purchasing estimating software.

Figure 7-2. Tips for Buying Software

- Consider that spreadsheet-oriented estimating systems provide instant, real-time results, so you can immediately understand the impact on the estimate. They also incorporate a database for storing and organizing information to reuse for each estimate. These systems typically offer the best of both generic spreadsheets and dedicated estimating systems.
- Look for a system that displays available options on where to go next either through function keys or a menu. Stay away from systems that require memorization of keys, such as "Alt" or "Ctrl" keys.
- Look for a window-based system, so that you can make an adjustment without searching through multiple screens. All takeoff work should be performed on the spreadsheet screen. If you have to leave the spreadsheet to perform takeoff, the system will be considerably less useful.
- Select a system that provides instant access to the database from within the estimate. If you must leave the estimate to edit your database, more often than not, you will not do it, and your database will become obsolete.
- Make sure the system you select is disk-based, not memory-based. A disk-based system automatically saves information while a memory-based system (i.e., generic spreadsheets) typically does not. Although disk-based systems are not as fast, they do ensure data integrity, and this capacity is far more important to an estimator than blazing speed.
- Select a system that incorporates a digitizer interface. A digitizer board on which the estimator places the drawings is an electronic tablet that enables the estimator to simply pick points on the drawings. The digitizer calculates the dimensions and areas automatically. This capability can be a tremendous time-saver during the takeoff process. You should make sure the digitizer interface displays the selected points on the screen because you must see them there. You should insist on this feature.
- Consider the hardware that will operate this estimating system. You will need a system with ample disk storage capability to handle your estimating program, a database, and a number of files. You should not forget about the other programs you will need, such as accounting, word processing, and scheduling. Processor speed is important, but not as important as disk speed or storage capacity. A math coprocessor will speed up your calculations, but only if the software takes advantage of it. A color monitor reduces eye strain, and many systems use color to identify different data. Finally, a high-quality printer will add to the professionalism of your work.
- Buy from a software vendor who provides the following:
  - A strong upgrade path (You may select a less-expensive base system with fewer features today, but you probably would like to upgrade in the future and receive full credit for your purchase.)
  - A competent support staff to answer questions within a reasonable response time
  - Software maintenance with annual or biannual updates
  - Modules for integrating CAD, scheduling, and other systems
  - An existing user base with satisfied customers

NAHB Software Review Program

The Business Management Committee of the National Association of Home Builders (NAHB) reviews construction-related business software.

To earn the NAHB-Approved Software® seal, the software must meet performance standards developed by builders. Software products are subjected to a series of controlled tests against those standards.

NAHB evaluators use the software, survey other users, and do on-site evaluation of the vendor's training and support capabilities. Only if the software meets NAHB's rigorous testing criteria does it earn, and have the right to use, the trademarked seal of approval in Figure 7-3.

Figure 7-3. NAHB Approved Software® Seal

NAHB Approved Product Summaries[1] describes the software and includes the following items:

- A performance summary
- A checklist of features
- Hardware requirements
- Technical support available
- Sample reports
- Sample screens
- Sample menus

Appendix

Conversion Tables

Table 1. Concrete Quantities for Footings and Grade Beams

This table provides cubic yard per linear foot conversion factors for cast-in-place concrete footings and grade beams. The dimensions of the member can be indexed to quantities/linear foot (in cubic yards) and multiplied by the length of the member.

Cubic Yard per Linear Foot

Member width (inches)	Member depth (inches)						
	6	8	10	12	14	16	18
6	.0092						
8	.0123	.0165					
10	.0154	.0206	.0257				
12	.0185	.0247	.0309	.0370			
14	.0216	.0288	.0360	.0432	.0504		
16		.0392	.0412	.0494	.0576	.0658	
18			.0463	.0556	.0648	.0741	.0833
20			.0514	.0618	.0720	.0823	.0926
22				.0679	.0792	.0905	.1019
24				.0741	.0864	.0988	.1111
26					.0936	.1070	.1204
28						.1152	.1296
30						.1234	.1389
36						.1481	.1667

Waste = 5 percent (+ 5 percent if earthen formed footing)

Other factors can be calculated by multiplying the width (inches) by the depth (inches) by .0002572.

Table 2. Concrete Quantities for Slabs

This table provides square foot factors according to slab thickness that can be used to calculate the volume (in cubic yards) of concrete required. Divide the area required by the area covered to obtain volume.

One Cubic Yard of Concrete

Slab Thickness (inches)	Area Covered (square feet)
1	324
1¼	259
1½	216
1¾	185
2	162
2¼	144
2½	130
2¾	118
3	108
3¼	100
3½	93
3¾	86
4	81
4¼	76
4½	72
4¾	68

Table 3. Header Size

This table helps determine header sizes not shown on details. Select the building width and opening size and determine the member size. The length of the members is the opening width plus 8.

Header Size	Single-Story Building Width*			Two-Story Building Width*		
	22-24	26-28	30-32	22-24	26-28	30-32
2 ea. — 2 × 4	3′10″	3′7″	3′2″	2′ 3″	2′	—
2 ea. — 2 × 6	6′ 1″	5′7″	5′0″	3′ 6″	3′1″	2′9″
2 ea. — 2 × 8	8′	7′5″	6′7″	4′ 8″	4′1″	3′7″
2 ea. — 2 × 10	10′ 3″	9′5″	8′4″	5′11″	5′2″	4′7″
2 ea. — 2 × 12	12′ 5″	11′6″	10′2″	7′ 2″	6′4″	5′7″

The header span is the maximum opening size that can be spanned by a double member. This is for lumber with allowable bending stress of 1,500 psi. For lumber with allowable bending stress of 1,000 psi, reduce allowable span by approximately 20 percent.

*The width is the dimension from one exterior bearing wall to the other exterior bearing wall.

Table 4. Reinforcing Bars

This table provides the weight/linear foot for each size bar to convert to pricing units. The lap factors are for continuous bars.

Designation	Diameter (inches)	Weight #/foot	Lap (40-bar diam.) 20″ bar	30″ bar	Factor 40″ bar
2	0.25	0.167	.042	.027	.021
3	0.375	0.376	.063	.042	.031
4	0.50	0.668	.083	.055	.042
5	0.625	1.043	.104	.069	.052
6	0.750	1.502	.125	.083	.063
7	0.875	2.044	.145	.097	.073
8	1.00	2.670	.167	.111	.083

Table 5. Spacing Conversion

This table provides factors to determine the number of spaces required based on the designated spacing. The calculation will give the number of spaces required, and one (1) to obtain the number of framing members required. There is no allowance for waste, doubling of members or intersecting walls (for take-off of studs).

On-Center Spacing (inches)	Multiply Length by	Divide Length by
12	1	1
16	0.75	1.33
20	0.60	1.67
24	0.50	2.00
30	0.40	2.5
36	0.33	3

Table 6. Quantity Multipliers for Flooring, Siding, Wall Board

Multiply the net square footage of the area to be covered by the multiplier in the table to determine the *area* of material required.

Type	Size	Multipliers Laid Straight	Diagonal
S4S	1 × 4	1.15	1.18
	1 × 6	1.14	1.17
	1 × 8	1.12	1.15
Ship Lap	1 × 6	1.27	1.30
	1 × 8	1.21	1.24
T & G	1 × 4	1.30	1.33
	1 × 6	1.21	1.24
	1 × 8	1.17	1.20
		Exposure	**Multiplier**
Bevel—1" lap	1 × 6	4½	1.35
	1 × 8	6½	1.25
	1 × 10	8½	1.20
Bevel—Rabbited (Drop siding also)	1 × 6	5¼	1.15
	1 × 8	7¼	1.10
	1 × 10	9¼	1.08

Conversion Area to Board Feet

1 × 4 area × 3 = bd. ft.
1 × 6 area × 2 = bd. ft.
1 × 8 area × 1.5 = bd. ft.
1 × 10 area × 1.25 = bd. ft.

Table 7. Roof Rafter and Sheathing Multipliers

To calculate the length of a rafter, take ½ of the building width plus the horizontal dimension of the overhang and multiply by the rafter factor listed in the table. The square footage of the sheathing can be determined by using the area factor. The valley or hip members can be determined by multiplying the horizontal dimension by the hip/valley factor.

Roof Slope	Rafter Factor	Hip/Valley Rafter Factor	Area/ Sq. Ft.
2/12	1.0138	1.4240	1.014
3/12	1.0308	1.4359	1.031
4/12	1.0541	1.4530	1.054
5/12	1.0833	1.4741	1.083
6/12	1.1180	1.5000	1.118
7/12	1.1577	1.5296	1.158
8/12	1.2018	1.5634	1.202
9/12	1.2500	1.6006	1.250
10/12	1.3017	1.6415	1.302
11/12	1.3566	1.6851	1.356
12/12	1.4142	1.7320	1.414
14/12	1.537		1.537
16/12	1.667		1.667

Lumber is purchased in 2' lengths, therefore calculations must be rounded up to even footage. Plywood sheathing square footage must be divided by 32 to determine the number of sheets required.

Table 8. Spacing to Units

Brick ties and other items are specified according to horizontal and vertical spacing requirements. This table provides conversion of spacing requirements to the number of pieces when the square foot of area is known.

Vertical Space	Horizontal Space	#/Sq. Ft.	Sq. Ft./Tie
12	8	1.5	.667
12	12	1.0	1.00
16	16	.057	1.75
24	16	0.38	2.67
30	16	0.30	3.33
36	16	0.25	4.00
24	24	0.25	4.00
30	24	0.20	5.00
36	24	0.167	6.00

Purchase unit is 100 or 1,000 lots for most items.

If noted as number of ties per square foot, or number of square feet per tie, the calculation is:

The total square feet × number of ties per square feet or
the total square feet + number of square feet per tie

Table 9. Roof Covering Factors

Add the proper percentage of roof area to allow for waste.

Roof	Wood Shingle	Asphalt Shingle
Shed/gable	8%	3%
Hip	12%	8%
Intersecting	12%	8%

Wood Shingle Coverage

Exposure to weather	Bundles/ square foot
4″	3.6
4½″	3.2
5″	2.88
5½″	2.62
6″	2.40
6½″	2.22
7″	2.06

Roof Sheathing Allowances for Waste
Change area to board measure.

Material	% Allowance
1 × 6 square edge	10
1 × 6 tongued and grooved	20
1 × 6 shiplap	20
1 × 8 square edge	12

Table 10.
Concrete Quantities—Walls

This table provides cubic yard (and cubic foot) conversion factors for cast-in-place concrete walls. The square footage of the wall surface multiplied by the factor listed below will provide the volume of concrete required.

Wall thickness (inches)	Cubic Yards/ Sq. Ft.	Cubic Feet/ Sq. Ft.
3	.0092	.25
4	.0122	.33
6	.0185	.50
8	.0250	.67
10	.0310	.83
12	.0371	1.00
14	.0430	1.17
16	.0490	1.33
18	.0550	1.50
24	.0740	2.00
30	.0920	2.50
36	.1111	3.00
42	.1300	3.50
48	.1480	4.00

Waste = 5%

Table 11.
Concrete Masonry Units—Quantities

This table provides factors to convert net square feet to concrete blocks and mortar. The factors are based on block laid in running bond with 3/8 inch mortar joints. The factor multiplied by the net square footage of the wall will provide the quantities required.

Blocks	Blocks/Sq. Ft.	Mortar/Sq. Ft. (cubic feet)
4 × 8 × 16	1.125	0.70
6 × 8 × 16	1.125	0.75
12 × 8 × 16	1.125	0.80
8 × 8 × 8	2.25	0.76
8 × 4 × 16	4.52	1.161
8 × 8 × 16	1.125	0.77

Table 12.
Truss-Type Reinforcement—Quantities

Horizontal joint reinforcement is ordinarily specified as equally spaced. Determine the number of courses in the wall and subtract 1; divide the result by the factor listed in the table and round off to the next highest round number. Multiply the number by the length of the wall to determine the total linear footage of joint reinforcement required. The standard length of each truss is 10′, and a 6″-lap is required. Add 5% for waste if openings are not deducted.

Horizontal Reinforcing Spacing	8″-High Course
8″ o.c.	1
16″ o.c.	2
29″ o.c.	3
32″ o.c.	4
40″ o.c.	5
48″ o.c.	6

Table 13. Clay Brick Quantities

This table provides factors to convert square footage of wall to brick and mortar quantities. Many estimators use a 5% factor for loss due to shipping breaks and on-site cutting.

Face Size Clay Brick	Running Bond Net Brick/sq. ft.	Brick with 5% Waste	Cuft Mortar with 25% Waste
Modular brick 2⅔″ × 4″ × 8″	6.75	7.00	.0588
SCR brick 2⅔″ × 6″ × 12″	4.50	4.75	.0988
Roman brick 2″ × 4″ × 12″	6.00	6.30	.0813
Norman brick 2⅔″ × 4″ × 12″	6.77	7.10	.0638

Table 14. Concrete Block Fill—Quantities

This table provides the conversion factors to determine the amount of grout fill or loose insulation for concrete block cell and bond beam fill. The height of the wall multiplied by the factor for cell fill provides the quantity (in cubic feet and cubic yards) of fill required. The total length of the bond beam and lintels multiplied by the factor in the table for U-block fill factor provides the quantity of fill required.

	Horizontal Fill		Vertical Fill	
	cu. ft./ lin. ft.	cu. yd./ lin. ft.	cu. ft./ lin. ft.	cu. ft./ lin. ft.
Block (standard)				
6 × 8 × 16	.137	.0051	.137	.0051
8 × 8 × 16	.204	.0076	.024	.0076
12 × 8 × 16	.371	.0137	.371	.0137
U-Block				
6 × 8 × 16	.148	.0055	—	—
8 × 8 × 16	.216	.0080	—	—
12 × 8 × 16	.394	.0145	—	—

Table 15.
Paint Covering Capability

Material	Square Feet Per One Gallon 1 Coat	2 Coats	3 Coats
Aluminum Paint	600	425	
Brick Paint—White or Light Tints on Unsurfaced Walls	225	110	75
Brick Paint—Dark Tints on Unsurfaced Walls	290	145	95
Enamel Base Coat (Prepared)	425	240	165
Enamels	425	215	165
Flat Wall Paint, Dark Colors (on Smooth Finish)	725	365	240
Flat Wall Paint, White or Light Colors (on Rough Sand Finish)	475	265	190
Flat Wall Paint—Sand Float Finish	300	150	
Flat Wall Hard Finish	500	300	
Flat—White Undercoat	500	300	
Linseed Oil	600		
Lacquer	200-300		
Lacquer Sealer	250-300		
Liquid Filler	250-400		
Non-Grain Raising Stain	275-325		
Oil Stain	300-350		
Outside House Paint—White or Light Tints, Porous Woods	475	255	
Outside House Paint—White or Light Tints, Close Grained Woods	575	300	190
Outside House Paint—Dark Colors, Greys, Tans, etc. Porous Woods	575	300	215
Paint and Varnish Remover (1 gallon should remove about 200 sq. ft.)			
Pigment Oil Stain	350-400		
Rubbing Varnish	450-500		
Stain, Wood Tints	750		
Stain, Shingle (2 gals. to 1000 Shingles for dipping 1 coat. Brushing 1 coat after dipping, ½ gal.)			
Spirit Stain	250-300		
Varnish Stain	550	350	
Waterproof Paint	450	250	
Water Stain	350-400		

Caulking Compound: Coverage ½″ × ½″ Ribbon 77/in. ft./gallon.

Paint Requirements for Interiors

Room Perimeter (feet)	Walls 8′ ceiling (gals.)	8′6″ ceiling (gals.)	9′ ceiling (gals.)	9′6″ ceiling (gals.)	Paint for ceiling	Finish for floors
30	⅝	⅝	¾	¾	1 pt.	1 pt.
35	¾	¾	¾	⅞	1 qt.	1 pt.
40	⅞	⅞	⅞	1	1 qt.	1 pt.
45	⅞	1	1	1⅛	3 pt.	1 qt.
50	1	1⅛	1⅛	1¼	2 qt.	3 pt.
55	1⅛	1⅛	1¼	1¼	2 qt.	3 pt.
60	1¼	1¼	1⅜	1⅜	2 qt.	3 pt.
70	1⅜	1½	1½	1⅝	3 qt.	2 qt.
80	1½	1⅝	1¾	1⅞	1 gal.	5 pt.

Each window and frame requires ¼ pt.
Each door and frame requires ½ pt.

Table 16. Nail Quantities

This table provides quantities of nails required for fastening lumber on carpentry items. Sum the quantities of lumber by type (in 1000 board foot measure—M fbm) and index to the table to determine the quantity, by weight, of nails required.

Item	Unit	Size	Pounds/ unit
Framing			
sills, plates, studs, joist, rafters	M fbm	10d common	6
Exterior Sheathing			
1 × 4	M fbm	8d	48
1 × 6 matched	M fbm	8d	32
1 × 8	M fbm	8d	27
1 × 10	M fbm	8d	20
gyp bd.	100 sq. ft.	4d-½″	1.5
plywood	100 sq. ft.	6d (or ed)	1
Flooring			
1 × 3 softwood	M fbm	2½ brads	32
1 × 4 softwood			26
1 × 6 softwood			18
hardwood	M fbm	1½ flooring	12
		2¼ flooring	24
Interior Sheathing			
gyp wallboard	100 sq. ft.	5d ring shank	1.0
plywood paneling	100 sq. ft.	3d finish	12
matched wood	100 sq. ft.	6d finish	6
Roofing			
asphalt shingle	square	1″ galvanized	3.0
wood shingle	square	4d shingle	3.0
Casing and Base	100 lin ft.	6d casing	1.0
Finishing Boards	M fbm	8d finishing	0.25
Furring and Masonry	100 lin ft.	1″ masonry	1.0

Notes

1. An Overview of the Builder's Estimate

1. Jerry Householder, *Scheduling for Home Builders*, 2nd ed. (Washington, D.C.: Home Builder Press, National Association of Home Builders, 1990), 83 pp.

2. The Complete Estimate

1. Bob R. Witten, *How to Hire and Supervise Subcontractors* (Washington, D.C.: Home Builder Press, National Association of Home Builders, 1991).
2. *Builder's Guide to Contracts and Liability*, 2nd ed. (Washington, D.C.: Home Builder Press, National Association of Home Builders, 1990).
3. Emma Shinn, *Accounting and Financial Management*, 2nd. ed. (Washington, D.C.: Home Builder Press, National Association of Home Builders, 1988).

5. Accuracy in Estimating

1. Emma Shinn, *Accounting and Financial Management*, 2nd ed. (Washington, D.C.: Home Builder Press, National Association of Home Builders, 1988), pp. 57-83.

6. The Cost Control System

1. Emma Shinn, *Accounting and Financial Management*, 2nd ed. (Washington, D.C.: Home Builder Press, National Association of Home Builders, 1988), pp. 57-83.
2. Shinn, *Accounting and Financial Management*, pp. 57-83.

7. Computerized Estimating

1. *NAHB Approved Product Summaries* (Washington, D.C.: Business Management Committee, National Association of Home Builders, 1991).

Glossary

boards—Lumber cut less than 2 inches in nominal thickness and 2 or more inches in nominal width.

black-in—The point at which the house has felt on the roof and sheathing on the walls.

brick veneer—The nonstructural, outside facing of brickwork used to cover a wall built of other material.

checklist—A listing of takeoff, subcontract, and cost estimate items used to organize and arrange the takeoff and cost estimate.

check-off—The method of using a checklist that requires each item shown on the working drawing to be checked off as it is counted or measured for the quantity determination and taken off for the cost estimate.

contingency cost—A cost that is added on when total project conditions cannot be anticipated nor estimated. Without a contingency cost, these unanticipated costs are offset by either the margin or unanticipated cost savings in other areas.

contract documents—The agreement between the builder and the home buyer including conditions, working drawings and specifications, addenda, modifications, and any other papers of official agreement.

cost estimates—Quantity takeoff summaries for work items listed in appropriate pricing units for figuring cost extensions to complete the estimate of labor, material, and/or equipment costs required.

cost extensions—The mathematical computations required to convert the quantity takeoff into a cost estimate. It requires multiplying the unit costs by the quantities determined and adding the individual cost estimates for each category (such as labor, material, and equipment) on each cost estimate form.

cost overrun—The amount by which the actual cost of an item exceeds the cost estimated for that item.

cubic foot—A 1 × 1 × 1-foot unit of volume. (A cubic yard contains 27 cubic feet.)

detail—An enlarged drawing of a part of another drawing that indicates precisely the design, location, composition, and relation of the elements and materials shown.

dimension lumber—Lumber that is at least 2 inches up to but not including 5 inches in nominal thickness and 2 or more inches in nominal width. Dimension lumber may be classified as framing, joist, planks, rafters, studs, and small timbers, and typically they are 2 × 4s, 2 × 6s, 2 × 8s, 2 × 10s, and 2 × 12s.

elevation—A two-dimensional, geometrical graphic representation of a building or object on a vertical plane—a picture view included in the contract documents.

equipment and jobsite overruns—The cost incurred when a project's duration and/or the use of a piece of equipment exceeds the estimate.

equipment and jobsite overhead contingency cost—A cost added as a percentage of the total equipment and jobsite overhead costs for the project. Some estimators increase the estimated time the equipment will be used or figure the rental rates higher than they will be.

equipment quantity—Equipment use measured in short-term rental rates (daily or hourly) for equipment used for a limited time. Long-term rental rates (weekly or monthly) are used for equipment required for longer durations (possibly on multiple sites that share cost).

estimate summary—A list of the total labor, material, equipment, subcontract, and overhead costs. It often includes the computation of labor burden, sales tax, and markup or profit.

estimate worksheet(s)—A listing of measurements, computations, and material quantities determined from a set of plans. The worksheet must have a title block that includes the name of the project, the number of the page, and the type of work covered on that page. Each worksheet and the sequence of worksheets should be organized according to the checklist for each takeoff and estimate.

furr down—Build down.

furr out—Build out.

jobsite overhead—Those costs for supervision, temporary utilities and facilities, layout, building permits, bonds, and insurance that are listed item-by-item and segregated by method of payment, including those computed on duration of use and those figured as a percentage of direct cost estimates. The cost extensions include labor, miscellaneous costs (e.g., layout), materials, and equipment.

indirect costs—The cost of labor burden and sales tax. Units for measuring labor burden are usually a composite percentage of direct labor costs. The sales tax is computed as a percentage of total material costs.

labor contingency cost is figured in anticipation of varying productivity in the building of the house. Builders often adjust the standard unit costs applied to the listed quantities of work to account for jobsite conditions. Unpredictable conditions, such as weather, affect labor productivity.

labor cost overrun—The amount by which the labor cost exceeds the labor cost estimate. Such overruns result from inaccurately calculating the quantity of work, an optimistic estimate of crew productivity (higher actual unit cost), or increased labor rates.

labor quantity—An amount of labor measured on a unit-cost and an hourly cost basis. The units of purchase for unit-cost labor will equal the labor-only units used in subcontracted items, including square footage of the building, roof, finish, or number of pieces (masonry). For labor bought on an hourly basis, labor costs depend on labor productivity rates from historical cost records. These records are expressed in the units of measurement used in cost control.

linear foot—A line 1 foot in length.

lump sum—A total cost for all of the work or material required to complete a segment of work for a house.

markup—Markup is the amount of money included in the estimate in addition to the direct costs of subcontracts, material, labor, equipment, and jobsite overhead. Some builders use a percentage of the subtotal cost for calculating markup. Other builders use a method of annual general overhead compared to the annual volume of work expressed in dollars or number of projects.

material contingency cost—An amount of money included in the estimate in anticipation of some loss of materials. Standard practice includes a percentage for loss in converting measurements or quantities to pricing units or as an addition to the number of units to be purchased.

material cost overrun—The amount by which material cost exceeds the estimate of that cost. These overruns occur when a price escalates or a greater quantity of materials is purchased than the estimate allowed.

material quantity—An amount of material measured in the same units as the units of purchase from vendors, including the amount of an item, board feet, and/or number of pieces by size, length, square feet, linear feet, sheet, or square.

plan view—A depiction of a horizontal section of a building or object through the walls. This contract document shows such items as openings, recesses, projections, and columns.

price—The current valid cost of a particular material if the material is bought at the time of inquiry.

price escalation—An increase in the price of a material between the time of inquiry (estimate) and the time of purchase.

price-escalation contingency cost—An amount of money added to the estimate to cover potential price increases for materials, usually as a percentage of the total of prices used in the estimate.

pricing units—The units of measurement used to express the quantities of materials on the quantity takeoff. Unit costs must be in the same pricing units as the quantity takeoff.

quantity—The amount of material, labor, equipment, or item of work required for one cost category of a house.

quantity determination—The measurement of individual items from the working drawings and computations required to determine a quantity of material or work required.

quantity takeoff—An organized arrangement and listing of quantity determinations arrived at by reviewing plans and specifications and measuring quantities.

quotation—A guaranteed price or bid, often with a time limit, offered by a vendor or subcontractor for a specific material or segment of work. Quotes are preferred because they are not subject to escalation and do not require contingencies.

schedule—A timetable for completing a job; also a list or table of parts including doors, windows, and room finishes.

section view—A drawing of an object as if it were cut lengthwise to show the interior makeup; often a contract document.

square foot—A 1×1-foot unit of area.

studs—A series of slender wood or steel members (typically 2×4s or 2×6s) used for the structural and nonstructural walls and partitions.

subcontract bid form—An organized format for recording and comparing subcontract bids and quotes. A similar form can be used to compare material quotes.

subcontract contingency cost—A cost that is added to lump-sum subcontract bids only when a bid is perceived or validated to be too low. Unit-cost subcontracts require builders to determine the actual billed units in their computations.

subcontract cost overruns—The amount of extra cost incurred for unit-cost contracts when estimated units do not equal actual measured units. When a subcontractor defaults on a bid, the unit or lump-sum cost of the work often exceeds the amount included in the cost estimate.

subcontract estimate form—A summary page of subcontract items formatted to list the subcontract bids and quotes used in the cost estimate. A similar form can be used to list and summarize the bids and quotes of material vendors.

timbers—Lumber that is nominally 5 or more inches in at least one dimension. Timber may be classified as beams, stringers, and posts, usually they are 6 × 6s and 8 × 8s.

unit costs—The estimated or quoted cost for a particular item of material or work per standard unit of measurement. The pricing unit used for cost must be the same as that used for the quantity takeoff.

Selected Bibliography

Builder's Guide to Contracts and Liability, 2nd ed. Washington, D.C.: Home Builder Press, National Association of Home Builders, 1990.

Clough, Richard H. *Construction Contracting*, 5th ed. New York: John Wiley and Sons, 1986.

Foster, Norman. *Construction Estimates From Take-off To Bid*, 2nd ed. New York: McGraw-Hill, 1973.

Householder, Jerry. *Scheduling for Home Builders, 2nd ed.* Washington, D.C.: Home Builder Press, National Association of Home Builders, 1990.

Means Residential Cost Data 1991. Kingston, Mass.: R. S. Means Co., 1990.

NAHB Approved Product Summaries. Washington, D.C.: Business Management Committee, National Association of Home Builders, 1991.

National Construction Estimator. Carlsbad, Calif.: Craftsman Book Co., 1991.

National Repair and Remodeling Estimator. Carlsbad, Calif.: Craftsman Book Co., 1991.

Park, William R. *Construction Bidding for Profit*. New York: John Wiley & Sons, 1979.

Peurifoy, Robert Leroy. *Estimating Construction Costs*. New York: McGraw-Hill, 1975.

Peurifoy, Robert L., and Gary Oberlander. *Estimating Construction Costs, 4th ed.* Construction Engineering and Project Management Series. New York: McGraw-Hill, 1989.

Shinn, Emma. *Accounting and Financial Management*, 2nd ed. Washington, D.C.: Home Builder Press, National Association of Home Builders, 1988.

Steinberg, Joseph and Martin Stempel. *Estimating for the Building Trades*. Chicago: American Technical Society, 1978.

Stewart, Rodney D. Cost Estimating. New York: John Wiley & Sons, 1982.

Tumblin, C. R. *Construction Cost Estimates*. New York: John Wiley & Sons, 1980.

Walker's Building Estimator's Reference Book, 23rd ed. Ed by Robert Siddens, et al. Lisle, Ill.: Frank R. Walker Co., 1989

Whitten, Bob R. *How to Hire and Supervise Subcontractors*. Washington, D.C.: Home Builder Press, National Association of Home Builders, 1991.